Chemistry of Sustainable Energy

Chemistry of Sustainable Energy

Rainer Roldan Fiscal

ARCLER
PRESS

www.arclerpress.com

Chemistry of Sustainable Energy

Rainer Roldan Fiscal

Arcler Press
224 Shoreacres Road
Burlington, ON L7L 2H2
Canada
www.arclerpress.com
Email: orders@arclereducation.com

ISBN: 978-1-77407-712-2 (Hardcover)

ABOUT THE AUTHOR

Rainer Roldan Fiscal is an Associate Professor I in Laguna State Polytechnic University, Philippines. He finished Bachelor of Secondary Education major in General Science, Master of Arts in Teaching Science and Technology and Doctor of Philosophy in Education major in Educational Leadership and Management. He teaches Biological and Physical Science subjects and Advanced Methods of Research in the undergraduate and graduate programs. He presented his papers in national and international conferences in the Philippines and abroad like Malaysia, Thailand, and Vietnam. He published his papers in international journals and has citations.

TABLE OF CONTENTS

GLOSSARY

A

Abridged – (of a piece of writing) having been shortened.

Alkaline – Having the properties of an alkali, or containing alkali; having a pH greater than 7.

Anode – Is an electrode through which the conventional current enters into a polarized electrical device.

Autocatalytic – Single chemical reaction is said to be autocatalytic if one of the reaction products is also a catalyst for the same or a coupled reaction.

B

Bacteria – Microscopic, single-celled organisms that thrive in diverse environments.

Biodiesel – A form of diesel fuel derived from plants or animals and consisting of long-chain fatty acid esters.

Biofuels – A renewable energy source, made from organic matter or wastes, that can play a valuable role in reducing carbon dioxide emissions.

Bituminous – Of, containing, or of the nature of bitumen.

C

Carbonates – A carbonate is a salt of carbonic acid, characterized by the presence of the carbonate ion, a polyatomic ion with the formula of CO_3^{2-}.

Catalysis – Is the process of increasing the rate of a chemical reaction by adding a substance known as a catalyst, which is not consumed in the catalyzed reaction and can continue to act repeatedly?

Catalysts – Is a substance that can be added to a reaction to increase the reaction rate without getting consumed in the process.

Catastrophic – Involving or causing sudden great damage or suffering.

Cathode – Is the electrode from which a conventional current leaf a polarized electrical device.

Chemisorption – Is a kind of adsorption which involves a chemical reaction between the surface and the adsorbate.

Circuit – Is an interconnection of electrical components or a model of such an interconnection, consisting of electrical elements.

Contaminants – Is the presence of a constituent, impurity, or some other undesirable element that spoils, corrupts, infects, makes unfit, or makes inferior a material, physical body, natural environment, workplace, etc.

Conventional – Based on or in accordance with what is generally done or believed.

Crevices – A narrow opening or fissure, especially in a rock or wall.

D

Decay – The state or process of rotting or decomposition.

Decomposition – The state or process of rotting; decay.

Displacement – The action of moving something from its place or position.

Disruption – Disturbance or problems which interrupt an event, activity, or process.

Dramatically – In a way that relates to drama or the performance of drama.

E

Electrochemical Cell – Is a device capable of either generating electrical energy from chemical reactions or using electrical energy to cause chemical reactions.

Electrode – Is an electrical conductor used to make contact with a nonmetallic part of a circuit.

Electrolyte – Is a substance that produces an electrically conducting solution when dissolved in a polar solvent, such as water.

Equilibrium – A state of rest or balance due to the equal action of opposing forces.

Eradicating – Destroy completely; put an end to.

Esterification – Is the process of combining an organic acid (RCOOH) with an alcohol (ROH) to form an ester (RCOOR) and water; or a chemical reaction resulting in the formation of at least one ester product.

Exaggerate – Represent (something) as being larger, better, or worse than it really is.

F

Fracking – The process of injecting liquid at high pressure into subterranean rocks, boreholes, etc., so as to force open existing fissures and extract oil or gas.

H

Heterogeneous – Is a somewhat comparative word, suggesting that two or more things are unlike in substance or nature, as in a heterogeneous mixture containing two substances that do not totally combine, like oil and water.

Homogeneous – Denoting a process involving substances in the same phase (solid, liquid, or gaseous).

Hydrate – A hydrate is a substance that contains water or its constituent elements.

Hydraulic – Denoting or relating to a liquid moving in a confined space under pressure.

Hydrocarbon – A hydrocarbon is an organic compound consisting entirely of hydrogen and carbon.

Hydrolysis – Hydrolysis is any chemical reaction in which a molecule of water ruptures one or more chemical bonds.

I

In Vitro – Vitro studies are performed with microorganisms, cells, or biological molecules outside their normal biological context.

Intermittency – Is the irregular alternation of phases of apparently periodic and chaotic dynamics, or different forms of chaotic dynamics.

Internal Combustion Machines – Is a heat engine in which the combustion of a fuel occurs with an oxidizer in a combustion chamber that is an integral part of the working fluid flow circuit.

Isomerization – Is the process by which one molecule is transformed into another molecule which has exactly the same atoms, but the atoms have a different arrangement, e.g., A-B-C → B-A-C.

L

Lipase – A lipase is any enzyme that catalyzes the hydrolysis of fats. Lipases are a subclass of the esterase.

Liquefaction – Is a process that generates a liquid from a solid or a gas or that generates a non-liquid phase which behaves in accordance with fluid dynamics.

M

Membrane – Is a selective barrier; it allows some things to pass through but stops others. Such things may be molecules, ions, or other small particles.

Mounting-A backing, setting, or support for something.

Mutagenesis – Mutagenesis is a process by which the genetic information of an organism is changed, resulting in a mutation.

O

Oxidation – Is the loss of electrons during a reaction by a molecule, atom or ion and it occurs when the oxidation state of a molecule, atom or ion is increased.

Oxidized – Undergo or cause to undergo a reaction in which electrons are lost to another species.

P

Permeability-The state or quality of being permeable.

Polymerization – Is a process of reacting monomer molecules together in a chemical reaction to form polymer chains or three-dimensional networks.

Polytetrafluoroethylene – Is a synthetic fluoropolymer of tetrafluoroethylene that has numerous applications.

Preservation – The action of preserving something.

Prolong – Extend the duration of.

Q

Quadrupling – Increase or be increased fourfold.

R

Radiative – Involving or accompanied by the emission of radiation.

Renewable energy – Energy is energy that is collected from renewable resources, which are naturally replenished on a human timescale, such as sunlight, wind, rain, tides, waves, and geothermal heat.

S

Sawdust – Powdery particles of wood produced by sawing.

Semiconductor – Semiconductor material has an electrical conductivity value falling between that of a conductor, such as metallic copper, and an insulator, such as glass.

Sewage – A type of wastewater that is produced by a community of people.

Sluggish – Slow-moving or inactive.

Solidifying – Make or become hard or solid.

Stimulation – Encouragement of something to make it develop or become more active.

Stripper – A device used for stripping something.

Surge – A sudden powerful forward or upward movement, especially by a crowd or by a natural force such as the tide.

T

Teeming – Be full of or swarming with.

Transmission – The action or process of transmitting something or the state of being transmitted.

Triggering – (of an event, circumstance, etc.) causing a particular action, process, or situation to happen.

U

Utility Grid – Is the utility's transmission and distribution system, that part of the system that transmits and distributes power to the customer.

V

Voids – An unfilled space in a wall, building, or other structure.

Z

Zeolitic – Zeolites are microporous, aluminosilicate minerals commonly used as commercial adsorbents and catalysts.

LIST OF FIGURES

LIST OF TABLES

LIST OF ABBREVIATIONS

AC	alternating current
ACS	American Chemical Society
ATR	auto thermal reforming
BETO	bioenergy technologies office
BOS	balance of the system
Btu	British thermal unit
BWRs	boiling water reactors
C	carbon
cal	calorie
CCS	carbon capture and storage
CFCs	chlorofluorocarbons
CH_4	methane
CHS	chemical hydrogen storage
CIGS	copper indium gallium selenide
CO	carbon monoxide
CO_2	carbon dioxide
COS	carbon oxysulfide
CPOX	catalytic partial oxidation
DAF-4	Davy faraday microporous solid no. 4
DC	direct current
DOE	department of energy
DPG	distributed power generation
EDTA	ethylenediaminetetraacetic acid
EE	enantiomeric excess
EIA	energy information administration
EJ	exajoule
EPA	Environmental Protection Agency

EU	European Union
FBRs	fast breeder reactors
FCC	fluid catalytic cracking
GHG	greenhouse gas
Gtoe	gigatons of oil equivalent
GW	gigawatt
H_2	hydrogen
H_2O	water
H_2S	hydrogen sulfide
HCFCs	hydrochlorofluorocarbons
HFCs	hydrofluorocarbons
HTS	high temperature shift
IEA	International Energy Agency
IPCC	intergovernmental panel on climate change
IT	international table
J	Joule
kWh	Kilowatt-hour
LHV	lower heating value
LNG	liquefied
LTS	low temperature shift
MBtu	million Btu
MCA	multi criteria assessment
MCFC	molten carbonate fuel cell
MEK	methyl ethyl ketone
MOF	metal organic frameworks
MSW	municipal solid waste
Mtoe	megaton of oil equivalent
NEI	Nuclear Energy Institute
NGLs	natural gas liquids
NIRS	nuclear information and resource service
NMP	N-methyl pyrledene
NMR	nuclear magnetic resonance
NOx	nitrogen oxides
NPT	non-proliferation treaty

NREAP	National Renewable Energy Action Plan
NREL	National Renewable Energy Laboratory
O_2	oxygen
OECD	organization for economic co-operation and development
PACCAD	panicoid, arundinoid, chloridoid, centothecoid, aristidoid, and danthonioid
PEM	proton exchange membrane
PEMFC	proton exchange membrane fuel cells
PLA	polylactic acid
PO	propylene oxide
POX	partial oxidation
Pr Ox	preferential oxidation
PTChem	polish chemical society
PV	photovoltaic
PVMA	PV market alliance
PWRs	pressurized water reactors
quad	quadrillion Btu
RCO	reduced crude oil
SAPO	silico aluminophosphate
SAREP	Spanish Association of Renewable Energy Producers
Si / Al	silicon to aluminum
SI	system of units
SOFC	solid oxide fuel cell
SR	steam reforming
SRREN	Special Report on Renewable Energy Sources and Climate Change Mitigation
TCC	thermo for catalytic cracking
TMTACN	tri methyl triazo cyclononane
TWyr	terawatt-year
US NRC	United States Nuclear Regulatory Commission
VOCs	volatile organic compounds
WGS	water-gas shift
WNA	World Nuclear Association
WNN	World Nuclear News

$ZnO–Cr_2O_3$	zinc oxide-chromic oxide
ZSM-5	zeolite socony mobil-5

PREFACE

Energy is an important part of the developing world. The researchers and the professionals around the world make all the efforts to ensure that the energy being employed across various sectors is sustainable and in accordance with a healthy environment.

The need for sustainable energy is increasing day-by-day as the demand for energy rises and the pressure on the resources increases along with the decrease in their availability. The researchers want to study more and more about the sources of energy that can improve sustainability and can save the other non-renewable sources while making it easy for the environment to handle the impact of their usage.

The book 'Chemistry of Sustainable Energy' basically tries to bring the focus of the readers to the various chemical processes involved in the energy production from the sustainable resources and explains how the various chemicals in those resources react to generate energy.

The first chapter takes the readers through the introduction to the concept of energy including the various forms of energy, its main sources, and the units in which energy is measured for different scenarios. The readers can equip themselves with the introduction before they move further to dwell on other concepts.

The second chapter brings the focus of the readers to the formation of the fossils and their extraction process. This chapter focuses mainly on the role of fossils in the generation of oil, coal, and gas, talking about the types of fossil fuels found on the earth's surface. This chapter also takes the readers through the refining of crude oil.

The third chapter informs the readers about the catalysis process and the production of hydrogen, as it is a major form of sustainable energy. The chapter dwells upon the production of clean hydrogen and the extraction of hydrogen from fossil fuels.

Then, the fourth chapter throws light on the manner in which the fuel cells and the photovoltaic cells help in the generation of energy and its utilization. The fifth chapter talks about the way in which nuclear energy and biomass are contributing towards the energy generation and how they promise to be the sources of the future energy requirements.

The sixth chapter takes the readers through the role of biofuels in the production of energy and explains the various chemical processes involved in the processes, so that the readers know about it in depth. Then, the seventh chapter focuses on the concept of green chemistry and explains the readers the way in which it relates to the sustainable energy goals.

The eighth chapter informs the readers about the current situation of the energy sector at the global level and how it sees sustainable energy as a part of future. The book has been designed to meet the knowledge requirements of the readers with the utmost detail and deep insights. I hope that the book is appreciated by the readers for the content it provides and the kind of knowledge it spreads.

Chapter 1

Basic Overview of Energy

CONTENTS

Energy is the fundamental unit of every life cycle. This chapter discusses about the significance of energy. While explaining the qualitative definition of energy and the evolution of energy as a concept, this chapter discusses the various forms in which energy is considered. Then different sources of energy are explained in this chapter. Majorly, this chapter explains the concept of renewable and non-renewable energy. Also, the chapter discusses about the units and the conversion of these units from one form to another. This chapter explains the concept of sustainable energy and its importance in the future. Lastly, this chapter explains the economic significance of energy in the modern world.

1.1. INTRODUCTION

In all the life cycles, energy is involved, and it is very much important in the field of agriculture as much as it is important in all other productive activities. The need of energy is already shown by an elementary food chain such as crops requires energy from solar radiation in order to grow, harvesting also requires energy from the body of the human in order to work, and cooking also requires energy from biomass in a fire. In its turn, the food provides energy to the human body.

Intensifying the production of food for higher output per hectare, and any other kind of progression in agricultural production, imply some extra operations. All these operations need energy. For example, fertilizing, land preparation and cultivation, transport, irrigation, and processing of crops.

Tools as well as equipment are used in order to support these operations. The production of this also needs energy (metallurgical processes, in sawmills, workshops, and factories, etc.).

The primary changes in the field of agriculture, such as mechanization and what is known as the "green revolution," imply some important alterations with respect to energy. Mechanization is defined as a change of energy sources, and often it means a net increase of the utilization of energy.

High yield varieties have been provided by the green revolution. But these high yield varieties could also be known as low residue varieties (that is per unit of crop). It has to be noted that this is exactly the residue which matters as the source of an energy for large groups of people in the rural populations.

In addition to it, there are some other sectors of rural life that also needs energy. There are some of the examples such as space heating, water lifting, the provision of shelter, and the construction of schools, roads, and hospitals. Moreover, social life also requires energy for the purpose of entertainment, lighting, communication, etc. It has been observed that development often implies some additional energy, and also different kinds of energy, such as electricity.

It has to be noted that energy is a scarce resource that is it is limited at least for some groups of individuals in some of the places and, maybe, it is a scarce resource for the entire world. Then it is very important and necessary to have a sensible or rational use of energy for economic as well as environmental reasons. This is applied to the field of agriculture as much as it is applied to any other area of the economy.

The understanding of the role of energy is very important for the rational use of energy.

1.2. QUALITATIVE DEFINITION OF ENERGY

In most of the dictionaries, energy is defined as "the capacity to do work." This means that energy is a more abstract concept as compared to that of work. Of course, the definition of energy is correct, but this definition of energy is not complete.

Work is indeed an essential 'manifestation' of energy. It has been observed that the Industrial Revolution went into full swing in the late period of 18th century when breakthroughs were attained in transforming other kinds of energy into work. But it has to be kept in mind that work is not just the 'palpable' kind of energy.

Heat is the other form of energy. It has been observed that a lot of effort as well as expense are made by people of the society in order to remove heat from their houses as well as offices in the summer season and to bring the heat to them in the winter season.

Also, the radiation, for better or for worse, is the type of energy that is sensed by everyone. Therefore, the complete definition of energy is the following: Energy is defined as a property of matter that can be transformed into heat, work, or radiation.

Both the works as well as the heat are the processes that can be used to change the energy. A good comparison is that energy is just like the money in the bank. On the other hand, work, heat, and radiation are just like checks, cash, and money orders.

It has been observed that most of the dictionaries have cited the year 1599 as the date of the original recorded use of the term 'energy' in English.

But it took the other two and a half centuries for the meaning of energy to be understood properly and completely. Indeed, the exact definition of energy, along with that of entropy, is one of the biggest intellectual accomplishments of mankind.

The instructive or informative analogy can be drawn between the rather abstract idea of energy and the other more or less abstract but recognizable property of (living) matter that is the health. Health is considered as a God-given property that everyone has, to which every individual often pays less attention, until they actually lose it.

In the similar way, energy is a property that is possessed by human beings and others, that is both living matter as well as nonliving matter, to a greater extent or lesser extent. It has been observed that people are pretty much uninformed of its existence, though, until it is transformed into heat, work, or radiation.

The definition of energy that is defined above underlines the fact that the conversion of energy is very important and required for the utilization of energy. It can be seen that the quantity of energy that is available in the universe is continuous. The quantity of energy cannot be created, or it cannot be destroyed; it is just converted from one form to the other form. In other terms, it is preserved in all these transformations. Work and heat (and some kinds of radiation) are the types of energy that is required by the society.

In the present times, these are also some forms of the energy that is used in order to produce electric energy (or it can be called as electricity). Unfortunately, though, these forms of energy are not the primitive forms of energy. They are not the forms of the energy that are available easily on the planet.

The primitive as well as the less palpable types of energy-like gravitational, solar, chemical, and nuclear energy is required to be transferred to heat, work, and useful radiation. The whole issue of the availability of energy can be decreased to that of transformation of ample but less convenient forms of energy into scarce as well as more convenient forms of energy that is required by the society.

1.3. ENERGY: EVOLUTION OF A CONCEPT

It has been observed that the historians of science have probably put in writing more numbers of essays on the development of the idea of energy than on any other area or subject. This idea of energy has developed from those of the outdated or archaic fire, the more modern Vis viva (that is "living force"), which was dominant until the 19th century. And this force was continued well into the nineteenth century and in the present times; this force has a much narrower scope.

Aristotle (384–322 BC) developed the concept of 'fire' as one of four fundamental components of nature. Firstly, it was explained by Empedocles (490–430 BC), the other three elements were water, earth, and air. This view continued to be unchallenged for the following two thousand years. The largely apathetic and profoundly spiritual 'experts' of the dark as well as the Middle Ages did not seem to care to make it clear.

Ultimately, the age of Enlightenment had to come. And this age had come in the late period of 17th century. Then the investigation into the how's as well as why's of this planet was not just resurrected or revived but it was organized into what in the present times is known as the scientific method.

This permitted Leibniz (1646–1716) to champion the concept that the Vis viva (living force) of a body is its mass times the square of its speed (in the present times, it is known to be twice the kinetic energy of a body). In the following 150 years or so, the quickly increasing scientific community was effective in drawing a clear difference between the more abstract ideas of force and energy and the less abstract ideas of work and heat.

The thermometer was invented as early as in the year 1592 by Galileo Galilei (1564–1642) and then it was invented by Fahrenheit (1686–1736) in the early period of 18th century.

Firstly, the invention of thermometer helped in clarifying the difference between heat and temperature. The complete studies of heat were provided by Joseph Black (1728–1799). This study of Joseph Black then encouraged James Watt (1736–1819) to create the first modern steam engine. This invention boosted the Industrial Revolution through the 19th century and beyond it.

In its turn, this important technological advancement encouraged the scientific community to solve those laws that govern the transformation of heat to work. In the contrary to the initial human realization that mass is preserved in all earthly as well as heavenly phenomena, the fact that preservation of energy is an even more fundamental law of the universe did not become clear until the middle of the 19th century. At this time, there was the development of the science of thermodynamics.

The most important people in this fascinating story of simultaneous findings are the American-born Benjamin Thompson (1753–1814), the Englishmen Thomas Young (1773–1829) and James Joule (1818–1889), the Frenchman Séguin (1786–1875) and the Dane Ludvig Colding (1815–10) and the Germans Robert Mayer (1814–1878) and Hermann Helmholtz (1821–1894)

While it has been observed that Thomas Young is well known for demonstrating the wave-like character of light and also for his study on the elasticity of materials, often he is praised for being the first to make use of the word energy in its modern scientific sense in the year 1807. Thompson is better known as Count Rumford. He is an amazing personality and also, he is very famous among all the historians for his military as well as political adventures. This has helped him in earning nobility in both England and Bavaria. In addition to it, he also explained the nature of heat by demonstrating that it is a substance that is not a fluid-like, as it was commonly believed until then. This was explained by Thompson in the year 1804.

This realization as well as the study of interconversion between work and heat by Colding, Séguin, Mayer, and particularly Joule, in the period between 1839–1849, explained the relationship between work and heat. The relationship between heat and work was explained as two qualitatively different but on the same time quantitatively equal forms of energy. Ultimately, in the year 1847 the encouraged young Helmholtz generalized the principle of preservation or conservation of energy into a universal law of nature.

This law came to be called the First Law of Thermodynamics that everyone is aware of. In addition to it, it has been observed that there is some irony in these historical progresses: just before main social upheavals were to spread through most of the regions in Europe in the revolutions of 1848. It has to be noted that the collective efforts of the scientific community of Europe brought about one of the primary intellectual syntheses of all time.

1.4. ENERGY IN ITS VARIOUS FORMS: FIRST APPROXIMATION

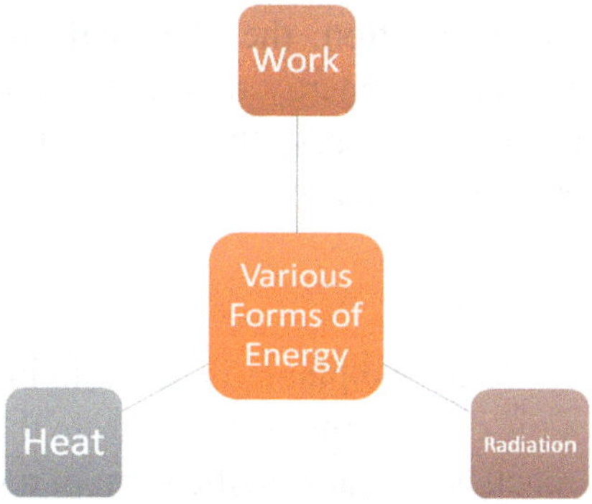

Figure 1.1. Major forms of energy.

1.4.1. Work

When an object is lifted (for example, the notebook) from one level to the other level (let's suppose that from the floor to a shelf on the wall), our energy is expended, by doing work, in order to increase the energy that is stored in the object (for instance, this energy can be transformed back into work as soon as the notebook is made to fall back to the floor) (Figure 1.1).

In this change that is discussed above, the chemical energy that is stored in the muscles of an individual is converted into work, or more precisely the energy is converted to mechanical energy, and work is transformed into the potential energy that is stored in the object (while the object is on the shelf).

1.4.2. Heat

When a kettle that is filled with water is kept on an electric stove, there is increase in the temperature of the water. This increase in the temperature is generally from 60°s F to 200°F. There is the increase in the energy of the water (sometimes it is known as the thermal energy), and coffee and sugar can be dissolved in it.

In this kind of transformation that is discussed above, electric energy is expended, and this energy is changed into heat. This in turn is shifted to the water. Let's suppose on a larger scale, the water is heated to a much higher temperature (let's say that 600°F) in an electric power plant (for instance, by the process of burning coal) and thus it needs sufficient amount of energy in order to produce electricity (by turning the turbines inside the magnetic field of an electric generator).

In this kind of transformation, the chemical energy that is stored in coal is transformed to thermal energy of water vapor. This energy in turn is transformed to mechanical energy (that is work) of the rotating turbine. Finally, this is converted into electricity.

1.4.3. Radiation

It has to be noted that life on the planet is possible just because of the Sun, and most of energy of the earth comes directly from the Sun. The energy that comes from the Sun reaches the earth in the form of a very wide spectrum of rays (or it can be called as waves) that are having different intensity, which is known as solar radiation. Most of the energy from the Sun is benign, such as radio waves and light (or it can be called as visible radiation). On the other hand, some of the energy from the Sun can also be harmful (for instance, ultraviolet rays).It has been observed that both evolution as well as technological progress has made the detection of this energy from the Sun possible. At one extreme, the first creatures that "have seen the light" existed on Earth some of the five hundred million years ago. On the other hand, the detection of the energy as well as the production of TV and FM waves is just few decades old.

In addition to the radiation that comes from the outer space, there are radiations that can also come from the smallest residents of matter. All matter (that is living matter and nonliving matter) is made of atoms. The protons as well as neutrons are in the nucleus and their electrons revolve around the nucleus.There is the storage of energy in atoms particularly the energy is stored in their nucleus (henceforth the name nuclear energy). When a nucleus (for instance, of uranium) is divided into small fragments, part of its energy is transformed into the required heat. On the other hand, the other part of the energy is transformed to different kinds of radiation and some of the radiations potentially have the harmful impacts.

1.5. SOURCES OF ENERGY

It has been observed that individuals have always made use of the energy in order to do work for them. It has been observed that thousands of years ago, early humans used to burn the wood to provide light, cook their food and heat the spaces where they live. It has also been observed that in the later years, individuals made use of the wind in order to move their boats from one place to the other place.

The individuals started making use of falling water, more than one hundred thirty-five years ago, in order to make electricity. In the present times, individuals utilized more energy than ever from a diversity of available sources for a multitude of tasks and the lives of the people are certainly better for it.

The homes of the individuals are comfortable, and these houses are full of valuable as well as entertaining electrical devices. The communication took place between individuals in many of the ways. Because of this, the individuals live healthier and longer lives. The people also travel the entire world, or at least they can see the whole world on television and on the internet.

In the present times, the ten primary energy sources that are used are categorized into two broad groups. These are nonrenewable sources of energy and renewable sources of energy.

1.5.1. Non-Renewable Energy

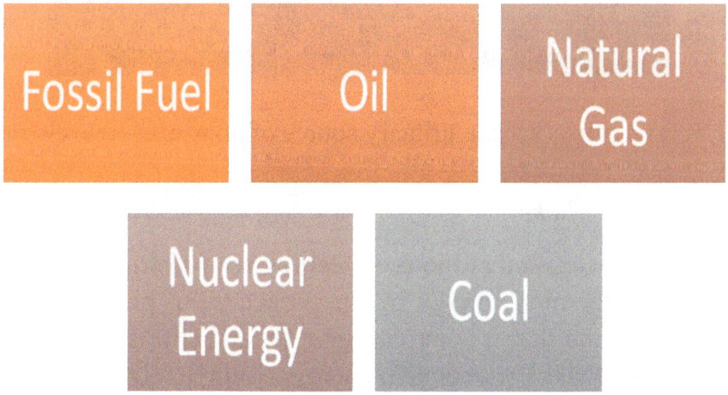

Figure 1.2. Different forms of non-renewable sources of energy.

Nonrenewable sources of energy consist of petroleum, natural gas, coal, propane, and uranium. These sources are used in order to produce electricity, to heat the homes, to move the vehicles, and to produce the products from candy bars to the mobile phones. All these energy sources are known as nonrenewable because these sources cannot be replenished in a short time period (Figure 1.2)

For instance, petroleum is a fossil fuel that was formed hundreds of millions of years ago, before the existence of dinosaurs. Petroleum was formed from the leftovers of ancient sea life, so it cannot be made rapidly. Also, it has to be noted that the world could run out of economically recoverable nonrenewable resources in the future.

1. **Fossil Fuels:** Fossil fuels (such as oil, coal, and natural gas) were formed from plants and animals. These plants and animals lived hundreds of millions of years before (before the existence of the dinosaurs). The fossil fuels were formed at the time of the Carboniferous period (Figure 1.3).

Figure 1.3. Fossil fuels are the primary source of non-renewable energy.

Source: Image by Flickr.

The plants that existed millions of years ago transformed the light energy of the Sun into chemical energy by the procedure of photosynthesis. This energy from the Sun that is solar energy was (and still is) moved down the food chain in animals. The chemical energy within the living organisms was trapped when living organisms die.

There are three important steps that are required for the formation of a fossil fuel. These steps are:

- Accumulation of organic matter (that is remains of animal or plant remains);
- Preservation of organic matter in order to prevent it from oxidizing (exclusion of air, for instance, by being in the swamp or the sea); and

- Conversion of organic matter into a fossil fuel like oil or natural gas.

Generally, this would happen because of the organic matter that is being shielded by layers of sediments. This increases the pressure as well as the heat (50–150°C). Fossil fuels are defined as the non-renewable sources because it takes a lot of years for this entire procedure to take place.

The fossil fuels burn and they produce carbon dioxide (CO_2). It is one of the greenhouse gases (GHGs). The burning of coal, this is also one of the fossil fuels, produces not just CO_2 but it also emits sulfur into the air. This emission of sulfur in the air increases air pollution.

2. Coal

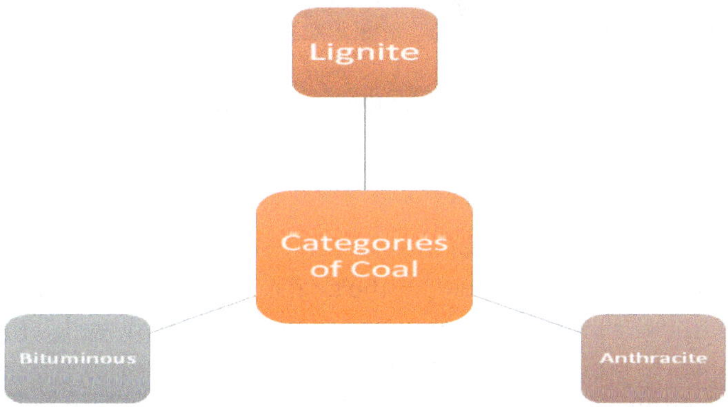

Figure 1.4. Different categories of coal.

Coal is defined as a solid form of fossil fuel that can be classified into three categories (Figure 1.4). These categories are

1. Lignite;
2. Bituminous; and
3. Anthracite.

It has to be noted that lignite coal is found close to the surface of the Earth. Thus, making it easy to mine, but lignite coal has high content of sulfur. Bituminous coal is defined as the most common coal that is burned, and this coal is less polluting as compared to that of lignite coal. The other category of coal that is anthracite is defined as the highest quality of coal. This coal is dark and shiny in nature and it is found deep inside the Earth.

Figure 1.5. Coal is still being used as the fuel in a number of industries.

Source: Image by Flickr.

Apart from all the pollutants from burning coal, coal mining creates many issues for the environment, as the coal must be mined from the ground. Huge quantities of unwanted dirt as well as rock are removed. This can give rise to the water pollution, unstable ground and, in most of the cases it has been observed that it is not appealing to look at. In addition to it, working in coal mines can be very risky or dangerous for an individual (Figure 1.5).

It has been observed that New Zealand has a number of coal reserves which includes the West Coast of the South Island and the Coromandel area.

3. **Oil:** Oil is a form of liquid fossil fuel. This can be yellow, dark brown or even green. It has to be noted that oil is very easy to mine once it is found. This is because being a liquid fossil fuel, oil will flow through pipes. This makes it very easy for transport. Though, oil can be difficult to locate. Oils are formed in reservoirs and in order to search for these reservoirs, the experts must study rocks as well as landforms to search and find the potential drilling sites.

After the hole is bored and if the oil is found there then the oil is piped to the surface. In this type, it is known as 'crude oil.' Crude oil is sent to a refinery. In refinery, this oil is heated up to different temperatures and different kinds of fuel are sorted out (for example, diesel, petrol, and jet-fuel) by the help of a process which is known as fractional distillation. The use of oil is not just for the purpose of transport, but it is also used in several different products like tires, plastics, and synthetic material like polyester.

4. **Natural Gas:** Natural gas is defined as a fossil fuel that is in the form of a gas (for instance, LPG, and methane). Often, natural gas is found beneath the oceans and near the deposits of oil. The process of performing the survey for natural gas reservoirs is same as that of oil exploration. The process of drilling is similar to oil once the natural gas field is found.

Natural gas is occurred deep under the surface of the earth. Natural gas mostly consists of methane (CH_4). Methane is defined as a compound which is having one carbon atom and four atoms of hydrogen. In addition to it, natural gas also consists of small quantities of non-hydrocarbon gases and hydrocarbon gas liquids. Natural gas is used as a fuel and is also used to make materials as well as chemicals.

It has been observed that millions of years ago, the leftovers of plants as well as animals (that is diatoms) decomposed and these were built up in thick layers, at times mixed with silt and sand. These layers were buried under silt, sand, and rock over time.

Some of this organic material was changed into coal, some of the materials into oil (petroleum), and some into natural gas by the pressure and heat. It has been observed that in some of the places, the natural gas was transferred into large cracks and spaces between the layers of overlying rock.On the other hand, in some other places, natural gas is occurred in the tiny pores (spaces) within some formations of sandstone, shale, and other kind of sedimentary rock. Here, this natural gas is mentioned as tight gas or shale gas. In addition to it, natural gas also occurs in the deposits of coal and is known as coal bed CH_4.

Gas can also be piped from its source and it can be stored for the use in the future. Natural gas is used for various purposes such as it is used for cooking and heating and apart from all of this it is also used for making a number of products like medicines, plastics, and fertilizers.

5. **Nuclear Energy:** Uranium-235 is used by nuclear energy. Uranium-235 is a radioactive chemical that is found naturally in the earth. This radioactive substance experiences decay and it releases a great deal of heat in this process. Then, this heat is used in order to create steam. This steam is further used to turn a turbine in a generator. As a result, the electricity is produced by the generator.

Figure 1.6. Nuclear energy is the most controversial form of energy.

Source: Image by Flickr.

In this process, no GHGs are released. It has to be noted that some sort of nuclear waste is produced by nuclear power plants which is known as radioactive elements. Strong radiations are emitted by these elements and these must be buried deep under the ground so that they don't have impact on the life of the humans. It has been observed that in the previous times, couple of nuclear disasters has already taken place which includes Chernobyl and Island Three Miles (Figure 1.6).

It has also been observed that in the recent past, there is the case of the disaster, which took place in Japan back in the year 2010. Again, these disasters have given rise to various questions on the security of nuclear power plants and those individuals who work in these plants.

Though, there are some radioactive byproducts that must be kept with all the safety measures. This is because these products are an environmental hazard. It is well thought-out as a nonrenewable form of energy. This is because it takes real manipulation to divide an atom.

1.5.2. Renewable Energy

These sources consist of solar, wind, hydropower, biomass, and geothermal. They are known as renewable energy sources. This is because the supplies of these energy sources are replenished in a short period of time.

The Sun shines day after day, the river flows and the wind blows. Renewable energy sources are particularly used for the production of electricity. Is electricity a renewable source of energy or nonrenewable source of energy? The answer to this question is neither. Electricity is different from the other sources of energy. This is because it is a secondary energy source. This means that the individuals have to make use of the other source of energy in order to make it. For example, in the United States, natural gas is considered as the number one fuel that is used for the energy and generation of electricity.

1. **Solar Energy:** The light from the Sun is one of the most abundant as well as freely available energy resources of the planet. The total amount of solar energy that reaches the surface of the earth in time span of one hour is more than the total requirements of energy of the planet for the entire year.

Even though solar energy sounds like a perfect renewable source of energy, the amount of solar energy that can be used differs in accordance to the time of day and it also differs according to the season of the year and the geographical location. For example, in the United Kingdom, solar energy is an increasingly common way to supplement the usage of energy.

2. Wind Energy

Figure 1.7. Wind energy is being used frequently in the modern times.

Source: Image by Wikipedia.

Wind is defined as an abundant source of fresh or clean energy. It has been observed that in many of the countries, wind farms are an increasingly familiar sight with wind power which is making an ever-increasing contribution to the National Grid. Turbines are used to drive generators in order to harness the electricity from the wind energy. This generator then feed electricity into the National Grid. Even though, there is the availability of domestic generation system or 'off-grid' generation systems, not every single property is appropriate for a domestic wind turbine (Figure 1.7).

3. **Hydro Energy:** Hydro power is a renewable resource of energy and one of resources that is most commercially developed. By the construction of a barrier, a large reservoir can be used in order to

build up a controlled flow of water that will drive a turbine, thus, producing electricity. It has to be noted that often, this source of energy can be more reliable as compared to that of wind power or solar (particularly if it is tidal rather than river).

In addition to it, this energy source also permits the electricity to be kept for the further use when demand of electricity reaches a peak. It has been observed that in certain circumstances, just like the wind energy, hydro can be more viable as a commercial source energy source (it depends on the kind and compared to other energy sources) but it is dependent very much on the kind of property, it can be used for domestic, 'off-grid' generation.

4. **Tidal Energy:** Tidal energy is the other form of hydro energy that makes use of twice-daily tidal currents in order to drive the turbine generators. However, tidal flow is not as constant as some other hydro energy sources, it is highly predictable. Thus, it can compensate for the periods when there is low tidal flow.
5. **Geothermal Energy:** Geothermal energy can be used to directly heat the homes directly or to produce the electricity. This is done by harnessing the natural heat below the surface of the earth. However, a power is harnessed directly below the feet, geothermal energy is not very important in many of the countries such as the United Kingdom as compared to the other countries such as Iceland. In Iceland, geothermal heat is available freely.
6. **Biomass Energy:** This is defined as the conversion of solid fuel into electricity. This solid fuel is made from plant materials. However, fundamentally, biomass includes the process of burning of organic materials in order to produce electricity and this is not burning the wood. And it has been observed that in the present times, this is a much cleaner and also, it is more energy-efficient procedure. By the conversion of industrial, agricultural as well as domestic waste into solid, liquid, and gas fuel, power is generated by biomass at a much lower economical cost as well as environmental cost.

1.6. ENERGY UNITS

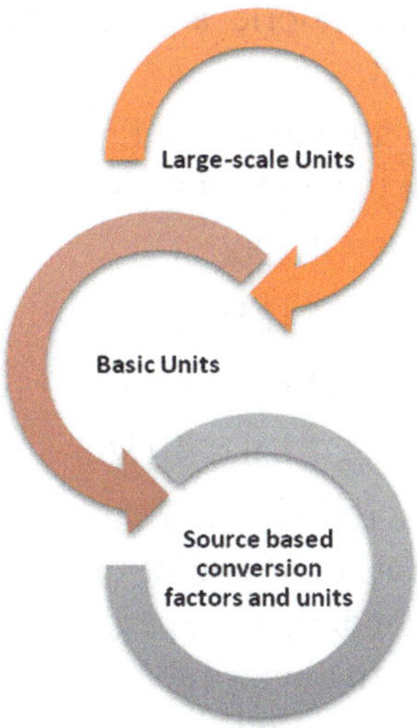

Figure 1.8. Different types of energy units.

There are various sorts of units that are used in the discussions of energy (Figure 1.8). They are categorized into two broad classes:

1. Those sorts of units whose definition is not associated to a specific fuel, which is termed as "basic" units; and
2. Those sorts of units whose definition is linked to idealized properties of a particular fuel, which is termed as "source-based" units.

1.6.1. Basic Units

1. **Joule (J):** Joule is defined as the basic unit of energy of the metric system, it is also known as the International System of Units (SI) for calculations and formulations that are easy to understand. Finally, it is defined in terms of the kilogram, second, and meter.
2. **Calorie (cal):** The calorie was described historically in terms of the heating of water. Therefore, defining it in a traditional way, one calorie is described as the amount of heat that is needed in order to increase the temperature of 1 gram of water by 1°C that is from the temperature of 14.5°C to 15.5°C. (Sometimes, this is also referred to as the 15°C calorie, and it slightly varies from the "calorie" measured for the other temperature intervals.) It has been observed that more recently the calorie has been described

in terms of the joule. It has to be noted that the equality between the joule and the calorie is historically called the mechanical equivalent of heat.

In addition to it, there are various definitions of the calorie that are in common use in the present times. This includes:

Mean calorie 1 cal = 4.1900 J

15°C calorie 1 cal = 4.1858 J

Thermochemical calorie 1 cal = 4.184 J (exact)

International Table calorie 1 cal = 4.1868 J (exact)

The publications of the Energy Information Administration of the U.S. Department of Energy (DOE/EIA) and of the International Energy Agency of the Organization for Economic Co-operation and Development (OECD/IEA) have adopted The International Table (IT) for their publications.

As the above-mentioned publications hold an important position and are referred by various other publications, it looks reasonable to view the IT calorie as the preferred unit for discussions of energy production and use. It must be noted that using IT calorie as the preferred unit for discussions of energy productions but there is no universally adopted practice for using the IT calorie as a preferred unit.

Also, it has been observed that in a few cases, a capitalized version known as Calorie is used to denote the kilocalorie (kcal). Usually while discussing food-related topics, the "calorie" is always mentioned. It must be noted that while mentioning the energy in food-related discussions, the word "calorie" always refers to kilocalorie.

3. **British Thermal Unit (Btu):** British thermal unit (Btu) is defined as the English system analog of the calorie. In order to express specific heat capacities in Btu/lb.-°F or in cal/gm-°C to be the same, the following conversion is used:

1 Btu = 251.9958 cal.

For the calorie, there is a family of Btus "Btu's" in relatively common use. This includes:

59°F (15°C) Btu 1 Btu = 1054.80J

Thermochemical Btu 1 Btu = 1054.35 J

Mean Btu 1 Btu = 1055.87J

International Table Btu 1 Btu = 1055.06 J

Again, it is the IT unit that is used in DOE publications or EIA publications.

4. Kilowatt-hour (kWh): The kilowatt-hour is defined as a standard unit of production of electricity and its consumption. By the definition of Kilowatt-hour (kWh), it has to be noted that 1000 watts = 1 kilowatt.

1 kWh = 3.6×10^6 J (exact).

The relationship between the Kilowatt-hour (kWh) and the Btu is dependent on which Btu is used. It is common, even though it is not universal, to make use of the equivalence:

3412 Btu = 1 kWh

This refers to the International Table Btu [in a more precise manner, 3412.14 Btu (IT) = 1 kWh].

1.6.2. Large-Scale Units

In explaining about the global energy budgets or national energy budgets, it is considered as a common practice to make use of large-scale units. These units are based upon the Btu, joule, and Kilowatt-hour (kWh):

1. Exajoule (EJ):

   1018 J = 1 EJ
2. Terawatt-year (TWyr):

1 TWyr = 8.76 x 1012 kWh = 31.54 EJ = 29.89 quad

3. Quadrillion Btu (quad):

1 quad = 1015 Btu = 1.055 EJ

1.6.3. Source-Based Conversion Factors and Units

In describing about the production and utilization of energy it is often considered suitable to speak in terms of the huge quantity of fuel, for example, a ton of coal or a barrel of oil. It has been observed that sometimes, these terms are used not just to denote a mass or a volume, but these terms are also used in order to represent the amount of energy.

1.6.3.1. Conversion Factors for Oil

The heat content of crude oil from various countries differs from about 5.6 million Btu (MBtu) per barrel to about 6.3 MBtu. The heat content of usual petroleum products has even more variations. A conversion factor, which might be negligent, is sometimes used for a barrel of crude oil. This conversion unit is close to its actual average energy content:

1 barrel of oil = 5.80 MBtu.

By using this definition, it is easy to establish a correspondence amongst millions of barrels of oil per day (Mbd) and quads per year:

1 Mbd = 0.0058 x 365 = 2.12 quad/yr

Usually, this is rounded off to: 1 Mbd = 2 quad/yr

Also, it is possible to establish energy equivalence for oil in terms of energy per metric ton (tonne). Unlike coal, the mass for oil is usually specified in tonnes. The number of barrels of crude oil per tonne varies widely, depending upon the source.

For 1993, the EIA has reported values varying from well under 7 barrels/tonne for some countries of origin to over 8 barrels/tonne for others. For the United States, the average was 7.33 barrels/tonne. This average, together with the nominal equivalence of 5.8 MBtu/bbl, corresponds to heat content for crude oil of 42.5 MBtu/tonne.

There are differing definitions in the literature of a tonne of oil equivalent (toe). In OECD/IEA publications it is set equal to 10.0 kcal (IT), while in other publications it is set equal to 10.7 x 10^6 kcal (thermochemical). These choices correspond, respectively, to:

1 toe = 1.00 x 1010 cal (IT) = 41.868 GJ = 39.68 MBtu (IT),

and

1 toe = 1.07 x 1010 cal (thermochemical) = 44.769 GJ = 42.46 MBtu (thermochemical).

In OECD/IEA tabulations, the megaton of oil equivalent (Mtoe), equal to 4.1868 x 10^{16} J, is used as the general unit to describe the energy content of all fuels. A corresponding larger unit, the gigatons of oil equivalent (Gtoe) can be related to the exajoule and quad:

1 Gtoe = 41.868 EJ = 39.68 quad.

1.6.3.2. Conversion Factors for Coal

Amounts of coal are described both in short tons and metric tons. Heat contents vary widely among countries, averaging less than 10 MBtu/ton for some regions and up to about 30 MBtu/ton for others, where low heat content corresponds to a large lignite fraction. For the United States in 1995, the average was 20.9 MBtu/ton.

In a fairly widely used specification of a nominal value, the energy content of 1 tonne of coal equivalent is set equal to 7 x 10^9 calories. Then, at a level of precision that makes the particular choice of "calorie" irrelevant:

1 tonne of coal (equiv) = 29.3 GJ = 27.8 MBtu

1 ton of coal (equiv) = 26.6 GJ = 25.2 MBtu

This nominal heat content is considerably higher than the average heat content of coal as presently consumed in the United States; it is, however, fairly close to the average heat content of coal used in the early 1950s.

1.6.3.3. Conversion Factors for Natural Gas

Natural gas is made up largely, but not entirely of CH_4 and its energy content is more uniform than that of coal. For the large majority of sources, the gross heat content of dry natural gas lies between 900 Btu/ft^3 and 1100 Btu/ft^3 (5). For the U.S. in 1995, it averaged 1028 Btu/ft^3.

There is no widely used equivalent unit of energy specifically based on natural gas, although the heat content of natural gas is often approximated by the rounded value of 1000 Btu/ft^3. In discussions of natural gas production and consumption it is also common to use the unit therm, where

1 therm = 100,000 Btu.

1.7. SUSTAINABLE ENERGY: FUTURE OF ENERGY

Over the last quarter-century, the twin concepts of sustainability and renewable energy have emerged as the factors that have entirely changed the entire human race. These factors have impacted every major sector such as science, technology, culture, economics, policy, and the environment. The twin concepts of sustainability and renewable energy are formulated in order to minimize the adverse effects of depleting natural resources, excessive energy consumption, water consumption and rapidly changing climate due to increase in the GHG emissions, which is a result of increased human activities.

As a result of the energy crisis, there have been several issues in environmental, economic, political, market, and social sectors of the society. Due to widespread issues and problems in the society, a number of researchers have been working in order to create or discover various sources that could provide sustainable and renewable energies. This will help in securing the energy consumption, conserving the environment, and will help in promoting development of less-developed regions. The implementation of sustainable and renewable energy is subject to various regions all over the world. This means that it totally depends on different communities how they integrate the sustainable and renewable records in their communities. In fact, various stakeholders have been more discreet and open about participating in processes and present their views, expectations, and framings in accordance with the major goals of introducing the sustainable and renewable energy sources.

There have been various methodologies and strategies that are generally used to address the issues related to sustainability and renewable energy. One such methodology, that is frequently used to address issues related to sustainable and renewable energy sources, is scenario planning. Scenario planning is one of the most effective strategies that is sought to address the issues. This is because scenario planning helps in formulating strategies that can put limits on uncertainty, while helping in improving the response capacity to a number of features.

Many scientists and developmental bodies, such as Kowalski et al. and Multi Criteria Assessment (MCA) used various combinations of scenario planning to mitigate the existence of high uncertainty in energy development. On the other hand, a number of stakeholders are involved in the decision-making process. The significance of involving a number of stakeholders is to involve a wide range of views on major social, economic, environmental, and technical aspects.

Simultaneously, many developmental bodies and researchers are majorly focusing on developing renewable energy as sustainable technologies. This is being done in order to focus on the associating various projects of renewable energy with the overall sustainability of a particular community or system. Many experts have stated that the renewable energy has contributed significantly to the sustainability of some particular regions and territories. The renewable energy provided these regions with a number of socioeconomic advantages and environmental benefits.

The assessments and evaluations made by the experts emphasize on the particular elements of sustainability. The major elements of sustainability that are in focus are the ecological, social, and economic elements. All these elements are emphasized with the help of monitoring and looking over the indicators. While monitoring these indicators, some cases are observed which majorly focus on the dynamic elements of sustainability.

Even though experts have varying opinions in regard with the potential of the enhancements used for increasing the efficiency in the energy sector and about the feasibility of CO_2 sequestration, almost every possible scenario indicates that only a notable expansion of the renewable energies provides the opportunity to the regions and communities to take up the path of energy sources and become sustainable.

Thus, it can be conferred that renewable energy is that only source that can be considered as a reliable source for fulfilling the supple needs of human beings in the future. One of the most distinguishing attributes of utilizing the renewable energy is that the lesser significant elements of the natural energy fluxes are withdrawn temporarily from the atmosphere. These lesser significant elements are returned later, in the form of lower-temperature heat. It must be noted that these minor elements of natural energy return to the atmosphere once the process of supplying the energy from the sources of renewable energy.

The Sun is the primary source of energy on planet Earth. The Sun is a source of energy that is inexhaustible. Most of the experts have explained the Sun as a huge reactor where fusion takes place continuously. This has been the main reason of the Sun supplying an unlimited, abundant, and safe energy to the planet earth for uncountable years and will continue for billions of years in the future. Every form of life that exists on the planet Earth derives its energy from the Sun. The entire flora on planet Earth grows and produces biomass only by utilizing the solar energy.

The Sun controls the weather, wind, and rainfall. Thus, it becomes indirectly responsible for creating the pre-conditions that give rise to wind power and hydropower. There are various kinds of devices that use the solar radiations without requiring any other device. Solar thermal collectors, photovoltaic (PV) systems, and concentrating solar power systems are some of these kinds of devices. Along with these devices, there are heat pumps that are capable of making the use of the ambient energy from the natural sources of energy such as air, water, and soil.

Another form of renewable source of energy that can be used is the geothermal energy. Geothermal energy is completely based on the radioactive decomposition that takes place in the inner layer of the Earth. After studying the entire human history, many experts have concluded that the human beings have majorly depended on the major renewable sources of energy. But most of the time, the human beings are using the renewable energy with very low rate of efficiency.

However, in the modern times, human beings have great advantages as compared to their historical counterparts. One of the biggest and the most significant advantage is the development of technology. Technologies developed in the modern world, and which are under development, make the utilization of renewable energy possible with high level of efficiency.

There are various kinds of associated costs with the renewable energy. But these associated costs are affordable if the renewable energy is used for a large-scale consumption. The only adverse scenario for renewable energy is if the chances for reducing the associated costs are exploited. Some of the commercial sectors try to exploit these chances. But this should be understood that these costs are always calculated in context with the environmental and social quality of the services provided. Also, there are external factors such as replacement of the usage of the fossil fuels and the impact on the entire fossil fuel industry.

After taking all these conditions into account, experts have found that the sources of renewable energy are capable of fulfilling all necessities of an energy supply system that is sustainable.

Most of the corporate firms have brought significant reforms and changes in their functional activities such as altering the distribution strategies and changing the packaging methods in order to become more sustainable and use energy responsibly.Most of the experts and researchers have stated that one of the most effective methods to become sustainable is by reducing the impacts of the way in which the energy is being consumed. This can be easily attained either by adopting the measure to save the energy or by using the renewable energy and become sustainable.

One of the most asked questions is why use renewable energy? In the modern times, the primary use of the fossil fuels is to provide heat and power to run the appliances at home and industries and provide fuel for the automobiles. Although, it is very easy and convenient to use the fossil fuels for fulfilling the energy needs, but there is a very limited supply of fossil fuels on the Earth.

Thus, the only sustainable source of energy is the renewable energy. Along with being replenishable, renewable energy pollutes less and, if effectively used, can be more efficient in comparison to the fossil fuels.

1.8. ECONOMIC SIGNIFICANCE OF ENERGY

Energy is a fundamental input to economic activity. Modern energy services light up our homes and schools, fuel economic activity to produce and consume, provide comfort and mobility, pump water and contribute to health and well-being. Harnessing energy sources to replace manual and animal labor was the platform of the Industrial Revolution: a period of unprecedented economic and social development.

The 20th century witnessed large increases in the global population, economic output and fossil fuel consumption. The gains from growth have been impressive for many. Yet these gains have taken a toll on a range of environmental systems where unsustainable practices have dominated. Continuing deterioration of natural resources could stress the ability to meet the needs of a growing population and undermine economic activity.

Green growth could meet this challenge. Green growth is about fostering economic growth and development while ensuring that natural assets continue to provide the ecosystem services on which our well-being relies. To do this it must catalyze investment and innovation which will underpin sustained growth and give rise to new economic opportunities. The energy sector poses a particular challenge in the context of green growth due to its size, complexity, path dependency, and reliance on long-lived assets. The current energy system is highly dependent on fossil fuels, whose combustion accounted for 84% of global GHG emissions in 2009.

Global demand for energy is rapidly increasing, due to population and economic growth, especially in large emerging countries, which will account for 90% of energy demand growth to 2035. At the same time, nearly 20% of the global population lack access to electricity. A major transformation is required in the way we produce, deliver, and consume energy. A large-scale transformation of the global energy sector is possible, though it will require significant investment. Global emissions could be halved by 2050, using existing and emerging technologies at an additional cumulative investment of USD 46 trillion.

It is vital for governments to create the enabling policy framework to catalyze private-sector investment in the transition to a low-carbon energy sector. It is cheaper in the long-term to act now, as every USD 1 of energy sector investment not spent before 2020 will require an additional USD 4.3 to be spent after 2020 to compensate for increased GHG emissions by building zero-carbon plans and infrastructure by the year 2035.

1.9. CONCLUSION

Energy is the most fundamental requirement for every life form and life function. The main question that has risen in the modern times is the replacement of non-renewable sources of energy with renewable sources of energy. All the experts and researchers are focusing in making the renewable sources of energy as the sustainable sources of energy. This is majorly because these renewable sources of energy are the only sources with which the human beings can survive without further degrading the environment.

REFERENCES

1. (2001). Renewable Energy: An Overview. [e-book] U.S. Department of Energy, p. 1. Available at: https://www.nrel.gov/docs/fy01osti/27955.pdf (accessed on 4 July 2020).
2. (2014). Renewable and Non-Renewable Energy. [e-book] Available at: https://www.ngemc.com/sites/ngemc/files/ERSY/LP%202.9%20Renewable%20and%20Nonrenewable%20Energy.pdf (accessed on 4 July 2020).
3. (2018). Introduction to Energy. [e-book] The NEED Project. Available at: https://www.need.org/Files/curriculum/infobook/IntroS.pdf (accessed on 4 July 2020).
4. Abdullah, M., (2017). Non-Renewable Resources that Cannot be Recycled Experiment Findings (PDF Available) January 2017 With 8,852 Reads (1st edn.). [e-book] Daffodil International University. Available at: https://www.researchgate.net/publication/312218812_Non-Renewable_Resources_that_cannot_be_recycled (accessed on 4 July 2020).
5. Altintas, K., Turk, T., & Vayvay, Ö., (2016). Renewable Energy for a Sustainable Future. Marmara Journal of Pure and Applied Sciences, [Online] 28(5). Available at: https://www.researchgate.net/publication/299433798_Renewable_Energy_for_a_Sustainable_Future?enrichId=rgreq-3dec44a6ef8d0f63f9f440322f445112-XXX&enrichSource=Y292ZXJQYWdlOzI5OTQzMzc5ODtBUzo0NDYzODAyMjIyMjY0MzJAMTQ4MzQzNjc0NjExMw%3D%3D&el=1_x_2&_esc=publicationCoverPdf (accessed on 4 July 2020).
6. Aps.org, (2020). Energy Units. [Online] Available at: https://www.aps.org/policy/reports/popa-reports/energy/units.cfm (accessed on 4 July 2020).
7. Davies, L., (2020). Types and Alternative Sources of Renewable Energy. [Online] EDF Energy. Available at: https://www.edfenergy.com/for-home/energywise/renewable-energy-sources (accessed on 4 July 2020).
8. Dürrschmidt, D., Zimmermann, G., & Liebing, A., (2004). Renewable Energies Innovation for the Future. [ebook] Berlin: Federal Ministry for the Environment, Nature Conservation and Nuclear Safety. Available at: https://www.dlr.de/tt/Portaldata/41/Resources/dokumente/institut/

system/publications/broschuere_ee_innov_zukunft_en.pdf (accessed on 4 July 2020).

9. Lund, H., Østergaard, P., Mathiesen, B., & Connolly, D., (2014). Renewable energy systems - A smart energy systems approach to the choice and modeling of 100 % renewable solutions. Chemical Engineering Transactions, [Online] 39, 1–6. Available at: https://www.researchgate.net/publication/289597903_Renewable_energy_systems_-_A_smart_energy_systems_approach_to_the_choice_and_modelling_of_100_renewable_solutions (accessed on 4 July 2020).
10. Mardani, A., Jusoh, A., Zavadskas, E., Cavallaro, F., & Khalifah, Z., (2015). Sustainable and Renewable Energy: An Overview of the Application of Multiple Criteria Decision Making Techniques and Approaches. Sustainability, [Online] 7(10), 13947–13984. Available at: https://www.mdpi.com/journal/sustainability (accessed on 4 July 2020).
11. Tam, C., Taylor, P., Girouard, N., & Konialis, E., (2011). OECD Green Growth Studies Energy. [ebook] OECD. Available at: https://www.oecd.org/greengrowth/greening-energy/49157219.pdf (accessed on 4 July 2020).
12. Walker, M., (1996). Chapter 2 Concept of Energy. [e-book] Available at: https://personal.ems.psu.edu/~radovic/Chapter2.pdf (accessed on 4 July 2020).

Chapter 2 Fossil Fuels: Formation and Extraction

CONTENTS

In the chapter formation and extraction of fossil fuel, various forms of fuels that come in the category of fossil fuel are discussed. The chapter also explains the types and formation of fossil fuels. It also tries to explain how fossil fuels are extracted from the ground. It also explains that fossil fuel extraction can also impact climate change by heating Earth from within.

In addition, it also throws some light on conventional vs. unconventional oil. The concept of shale oil and gas is also explained. The refining of crude oil is also discussed in this chapter. The various processes involved in the refining of crude oil are also discussed in this chapter. Some of these processes are fractional distillation, vacuum distillation, cracking, solvent treating, solvent dewaxing, and thermal cracking. Moreover, the refining of natural gas is also discussed. In the end, how carbon is captured and stored is discussed in detail.

2.1. INTRODUCTION

Fossil fuels are broader terms that entail hydrocarbons, primarily coal, natural gas, or fuel oil, created from the remains of dead animals and plants.

In a general way, the term fossil fuel also involves hydrocarbon comprising natural resources that are not derived from plant or animal sources.

Sometimes, these are also referred as mineral fuels.

The significant rise in the use of fossil fuels has made it possible for large-scale industrial development and chiefly supplanted water-driven mills, as well as the burning of peat or wood for heat.

In general, fossil fuel refers to a term for buried combustible geologic deposits especially of organic materials, produced from decayed animals and plants that have been transformed to coal, crude oil, heavy oils or natural gas by exposure to pressure and heat in the crust of earth over hundreds of millions of years.

It is important to note that human activity contributed heavily for the burning of fossil fuels which is the main cause of emissions of CO_2 that is responsible for forming various greenhouse gases (GHGs), allowing radiative forcing and contributing to global warming.

A trivial percentage of hydrocarbon-based fuels are biofuels originated from atmospheric CO_2, and thus do not escalate the total quantity of CO_2 in the atmosphere.

Fossil fuels, involving oil, coal, and natural gas, are presently the world's major energy source. As it is well known that fossil fuels are formed from organic material over the course of millions of years it triggered global and U.S. economic development over the past century.

Yet fossil fuels are of limited in nature and can cause significant harm to the environment that is impossible to rectify or repair. According to a report published by U.S. Energy Information Administration, it was found that the burning of fossil fuels was responsible for minimum 76% of U. S. total GHG emissions in the year 2016. These gases are primarily responsible for greenhouse effect and are majorly accountable for possibly catastrophic changes in the Earth's climate.

The recent advancement in technologies such as Carbon Capture and Storage (CCS) may help in lessening the GHG emissions generated by fossil fuels, and nuclear energy can be a zero-carbon substitute for generation of electricity. In addition, there are also various other sustainable means and less risky solutions available in the market such as: energy efficiency and renewable energy.

All these energies if produced in adequate amount will help in fulfilling the demand of current generation and further help in preservation of critical resources for future generations. If the present generation only relies on the fossil fuels and other nonrenewable energy, then these resources will not be available for future generation as it is limited in nature.

2.2. OIL

Oil is the world's primary fuel source for transportation. Without it, the entire transportation services all over the world will suffer. Major proportion of oil is extracted from underground reservoirs, but it can also be found embedded in tar and shale sands.

Once extracted from the available sources, crude oil can be processed in oil refineries in order to create fuel oil, liquefied petroleum gas, gasoline, and other nonfuel products such as fertilizers, pesticides, plastics, and pharmaceuticals.

At present, the United States is the dominant player in the consumption of petroleum products at 19.88 million barrels per day as of 2017. As the U.S. is one of the major producers of petroleum products, it also imports these to fulfill its current demand.

According to data, it imports around 3.8 million barrels per day. Top exporters to the United States include Mexico, Canada, Venezuela, Saudi Arabia, and Nigeria. As the entire world rely heavily on oil for transportation, it is very difficult to reduce its consumption.

But oil poses major environmental problems. Besides the environmental degradation caused by extraction of oil and oil spills, combustion of oil releases fine particulates that can cause significant respiratory problems. In addition, oil is also considered a major source of GHG emissions: petroleum products contributed for at least 45% of GHG emissions in the United States as of 2017.

Heavier crude oils, exclusively those extracted from tar sands and—through fracking—from shale, necessitate the use of energy intensive methods which resulted in causing more emissions and environmental degradation in comparison to conventional oil. As conventional oil from underground reservoirs runs out, various oil producers are turning towards unconventional sources such as oil shale and tar sands.

2.3. COAL

Coal plays a very important role for fulfilling the electricity demand of the entire world. According to a 2017 survey in the United States, it was found that coal is widely used to generate electricity and is responsible for 30% of the electric power supply (down from 39% in 2017 and 50% in 2007). The United States produces around 11% of the world's total with West Virginia, Wyoming, Illinois, Kentucky, and Pennsylvania leading in production. China is the dominant leader in production of coal, responsible for nearly 48% of world supply (as of 2015).

The combustion of coal is primarily responsible for discharging air pollutants such as acid rain-inducing nitrogen oxides (NOx), sulfur dioxide, and mercury.The mining process has the potential to significantly damage the environment, often resulted in havoc of vegetation and top-soil. Rivers and streams can also be contaminated or destroyed by mine wastes. It was found that the combustion of coal is accountable for 32% of the GHG emissions in the United States.

With the increase in awareness among people, government, and various environmental group the premise of "clean coal" has recently been promoted as a way to use this rich energy source without causing much damage to the environment.

CCS, that is, used for separation of carbon from coal and is injected underground to store for prolong use, could theoretically be used to alleviate the coal industry's GHG emissions.However, because of lack of research, CCS has not yet proved to be safe or realistic way to diminish GHG emissions from commercial power plants, and the health and environmental-related costs of mining remain.

2.4. NATURAL GAS

Natural gas contributes a significant portion of U.S. electricity and now exemplifies approximately a one-third of the country's energy use. It is most widely used for production of electricity or heat for industrial or buildings processes; less than three% of U.S. natural gas is used as a fuel for transportation activity, archetypally for bus fleets.

It was estimated that the United States produces roughly 17.6% of the world's natural gas and consumes about 21.6% of it (as of 2015). Natural gas is majorly supplied through pipeline that makes Canada the key exporter to the United States, while Russia is the dominant supplier to the European countries. Increasingly, to countries that located wide apart, natural gas is being transported by ship in a liquefied form (LNG) in order to fulfill the greater global demand for the fuel.

Natural gas is much cleaner when compared with oil and coal, with almost zero sulfur dioxide emissions and far fewer nitrogen oxide and particulate emissions. According to various surveys, it was found that natural gas discharges almost 30% less CO_2 compared oil and 43% less than coal. However, still natural gas is accountable for 29% of GHG emissions in the United States.Natural gas that is mainly composed of methane (CH_4) is also produced by the manure from production of livestock and decomposition of municipal waste in landfills.

CH_4 is a GHG that is much feasible than CO_2. Capturing and burning the gas to produce usable power and heat avert CH_4 from being discharged from the feedlot or landfill into the atmosphere directly.

2.5. TYPES AND FORMATION OF FOSSIL FUEL

Formation of fossil fuels refers to the process that takes place over the time span of around hundreds of millions of years to generate a wide category of fossil fuels including oil, coal, and natural gas. Although, it is generally seen that majority of the produced materials begin as organic material and are then altered over a large time span due to burial by pressure, sediment, and temperature, different environmental conditions result in the emergence of variety of fossil fuels.

Even within one "class" of fossil fuel (such as oil) there are a number of means that the oil can form and each method results in a distinct oil deposit, sometimes requiring special techniques to access.

2.5.1. Types of Fossil Fuel Formation

Figure 2.1. A particular category of fossil fuels.

Source: Image by Pxfuel.

There are basically three key types of fossil fuel formation and they are:

a) **Oil Formation**: Although conventional oil reserves form in fairly the same way, the oil sands produced in a slightly different way. Additionally, shale oil refers to oil that from a special type of reserve from kerogen trick inside low permeability shale (Figure 2.1)

b) **Natural Gas Formation**: Traditional natural gas reserves form in principally the same manner, however tight gas and shale gas are two chief types of natural gas. Tight gas is a gas that trapped inside a rock with exceptionally low permeability—usually sandstone or limestone. In contrast, shale gas refers to a category of natural gas that is trapped within shale formations.

c) **Coal Formation**: Formation of coal is completely different from that of oil and natural gas formation as it is buried for a prolong period of time, resulted in formation of different sorts of coal. Namely, extended burial results in the formation of sub-bituminous coal, lignite, anthracite, and bituminous coal.

2.5.2. Formation of Oil and Gas

It is generally seen that oil and gas are basically created from organic material primarily deposited as sediments on the seabed and then shattered in small amount and then transformed over millions of years. If there is an appropriate combination of reservoir rock, source rock, a trap, and caprock in an area, it is more likely that a recoverable oil and gas deposits may be discovered there.

Majority of the oil and gas deposits on the Norwegian shelf create from a thick layer of black clay that presently situated thousand meters beneath the seabed.

The black clay is a source rock that means it a deposit comprising large amount of organic residue. In a survey, it was found that the clay was deposited around 150 million years ago at the bottommost of a sea that covered much of current era northwestern Europe. Majority of the seabed situated here was stagnant and dead, while the upper water layers were teeming with life.

With the death of microscopic phytoplankton, they sank to the bottom and gathered in huge amount in the oxygen-free sediments. Over period of time, they were deeper buried and exposed to a long process of chemical conversion by decomposition of bacteria and ripening under a thickening pile of sediment. This resulted in the creation of gaseous and liquid hydrocarbons in the source rock.

One product that comes under the category of anaerobic decomposition of organic matter is kerogen which at high pressure and temperature gradually produces oil and gas. On the Norwegian continental shelf, the temperature upsurges by about 25°C per kilometer of depth.

After more than a hundred million years of sedimentation and erosion, the source rock may be buried under numerous kilometers of sand and clay deposits. There is formation of oil when the kerogen temperature reaches in between 60 to 120°C; at high temperatures, it is largely gas which is produced.

As oil and gas form, they flow or leak slowly through source rock. Because hydrocarbons are less in weight compared with water, the gas and oil travel upwards in porous water-bearing rock.

It is generally seen that the migration of oil and gas takes thousands of years and may lengthen over tens of kilometers until it becomes stationary by impermeable layers of rock, or the gas or oil leaks out into the sea.

Reservoir rocks are porous and always saturated with oil, water, and gas in several combinations. Majority of the petroleum products are trapped in reservoir rocks accumulated in large deltas created by rivers that ran into the sea during the Jurassic Period.The main reservoirs of the Oseberg, Gullfaks, and Statfjord fields are in the large Brent delta that shaped in the Jurassic. There are also hefty reservoirs located in sand that was consigned on alluvial plains during the time of Triassic Period (the Snorre field), in little depth seas in the Late Jurassic (the Troll field) and as subsea fans during the Paleogene Period (the Balder field).It was found that in the southern part of the North Sea, thick layers of chalk constituted of microscopic calcareous skeletons of animals and plants build an imperative reservoir rock, as in the Ekofisk Field.Mudrocks and other impermeable deposits inspire immigration routes from the source rock to the reservoir.Moreover, it is essential to have impermeable rock in order to prevent petroleum escaping from reservoir rock. Impermeable rock that comprises a seal over reservoir rocks referred as called cap rock. In addition, the arrangement of the reservoir rocks should be in such a way that the oil accumulates in a trap.In case there is adequate combination of reservoir rock, source rock, a trap and cap rock in an area, then it is possible to discover the recoverable oil and gas deposits there.

2.6. HOW ARE FOSSIL FUELS EXTRACTED FROM THE GROUND?

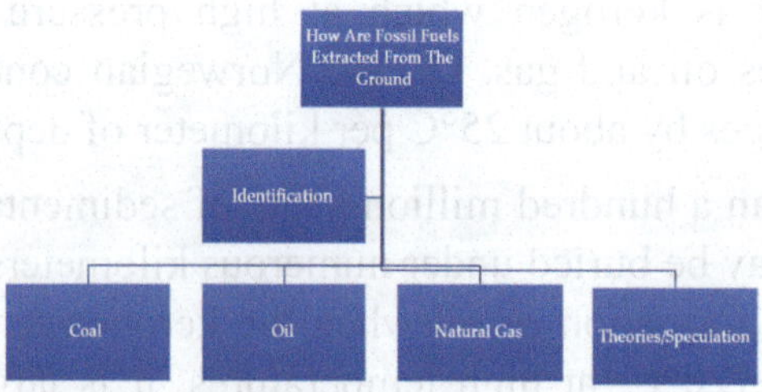

Figure 2.2. How are fossil fuels extracted from the ground?

It is almost impossible to run economy today without the use of fossil fuel. There are basically three categories of fossil fuels: natural gas, coal, and petroleum (oil). Fossil fuels are primarily generated from dead organic matter millions of years ago. The present perception is that the society relies heavily on fossil fuels, which is the critical factor of current environmental and public health crisis (Figures 2.2, 2.3).

Figure 2.3. Machinery used in extraction of fossil fuels.

Source: Image by Pxfuel.

2.6.1. Identification

Fossil fuels generate come from animal and plant matter that died thousands of millions of years ago. Soil and sediment take numerous years to build up, placing pressure on the material and forcing oxygen out. This plant matter then turned into kerogen that transformed into oil when its temperature rises up to 110°C. In a similar way, natural gas can be formed from oil at temperatures above 110°C.

2.6.2. Coal

It is worth noticing that the most of all mining for fossil fuels includes the extraction of coal. Coal can be taken out from upper portion of the earth's crust, known as surface mining, or from deep within the Earth with the help of underground mining. Recovery of coal through surface mining is comparatively easy; bulldozers and shovels are effective at extracting coal close the surface. Once exhausted, workers replant a surface mine and move on.

2.6.3. Oil

Onshore oil derricks and offshore oil rigs drive most of the petroleum that is obtained throughout the world. This process is usually done by drilling a hole into a potential oil patch and then oil is propelled out via a long tube. According to the EIA, in the United States, some dominant oil producing states are located along the coast.

2.6.4. Natural Gas

It is often seen that petroleum products and natural gas are often found in the same patch of land. Scientists look for oil and gas deposits with exceptional tool that causes a vibration in the ground as assured regularities are associated with oil and gas.Pumps then separate gas and oil on site. New technology, called "digesters," allows generating natural gas from plant matter by triggering and zooming up the natural process.

2.6.5. Theories/Speculation

The Environmental Protection Agency (EPA) is of the opinion that the burning of fossil fuels is majorly responsible for global warming.

Fossil fuels when burnt release CO_2, a gas that traps heat below the earth's atmosphere level, resulting in having a greenhouse impact. According to a recent research, the world releases large amount of CO_2 that could warm the earth in a comparatively shorter period of time.

2.7. FOSSIL FUEL EXTRACTION COULD BE CONTRIBUTING TO CLIMATE CHANGE BY HEATING EARTH FROM WITHIN

It was noticed that majority of scientists agreed to the point of view burning fossil fuels is contributing to climate change. But is no clear notion behind the fact that how exactly it's playing role in rising global temperatures.

According to various data, the world is now 1 °C warmer if one compared it with pre-industrial times. So, is it right to assume that it is just due to the emissions of greenhouse gasses like carbon dioxide? Meteorologist Hubert Lamb is considered as the father of modern climatology. He is of the belief that CO_2 levels couldn't alone accountable for the current global warming conditions that's been observed.

His major area of focus is rather on the role of thermal emissions. Burning of fossil fuels cannot only direct to release of GHGs, it also discharges a lot of heat, which come in contact of atmosphere. Some other examples of large heat sources are nuclear tests and volcanic eruptions.

Back in the year 2009, two scientists of Sweden region maintained that thermal emissions were more important than CO_2 for growing global temperatures level. A few years later, two Chinese scientists proposed that heat from the Earth's interior could be major factor in surging temperature level.

They are of the belief that fossil fuels such as oil, coal, and gas in crevices and layers beneath the Earth's surface act as a protecting blanket, trapping heat from the planet's interior. As these deposits have been drained by extraction of fossil fuel, more of that heat could be reaching the surface.

This notion is behind the concept of how fat tissue under the skin prevents body heat from being lost to the surrounding air. To examine this theory in the Earth's crust, it is important to study some valuable information and data for global production of fossil fuel alongside data for variation in temperature on the sea and land surface.

The present available research demonstrated that it is possible that temperatures may be rising faster in places where fossil fuels are being extorted from the ground (Figure 2.4).

2.7.1. Rising Heat

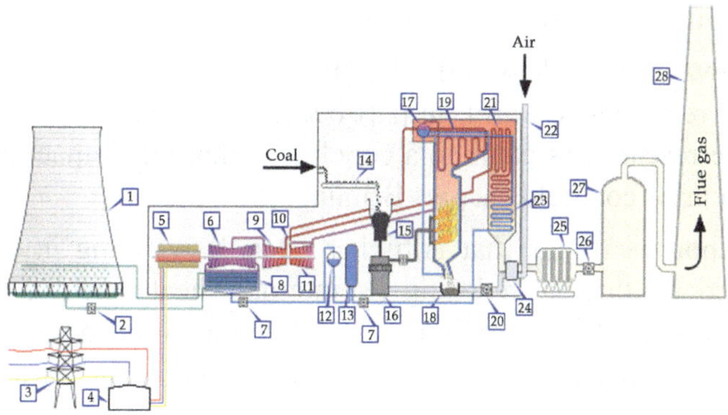

Figure 2.4. Thermal power station.

Source: *Image by Electricalvoice.*

Between 2007 and 2017, around 45.5 billion tons of oil and 36.3 billion cubic meters of natural gas were apart from the Earth's crust. When there is extraction of gas and oil, the empty space fills with voids water, which is a less effective insulator. This means more heat from the interior of Earth can be transferred to the surface, causing the ocean and the land to warm.

It is therefore essential to take a look at the warming trends in gas and oil producing regions all over the world. These places, which included the Arabian Gulf, Saudi Arabia, the North Sea, Gulf of Mexico, and Alaska, witnessed high rates of warming-which is almost three to six times higher compared with the average rate worldwide.

One of the fastest rates of warming has been witnessed in the Arctic, where the rise of temperatures observed by 0.6°C every decade since 1978. However, in Antarctica, the increase is just 0.1°C, despite similar levels of atmospheric CO_2 in both Polar Regions.One major reason for this variation could be because of the fact that fossil fuels are extracted in the Arctic, but not in the Antarctic. From 2007, more than 400 gas and oil arenas have been developed north of the Arctic Circle, while in Antarctica the extraction of fossil fuel is completely banned.

In an earlier study, a similar pattern has been observed in the northeast of England, where a long history of coal mining has considerably changed the lands subsurface. So much that in the prior coalfields around Newcastle and Gateshead, a "heat island" outcome was noticed beneath and below the ground. This meant the atmosphere above the conurbation was around 2°C warmer compared with surrounding area, while the ground underneath Gateshead was noticed to be around 4.5°C warmer.

Groundwater that released from a mine water pumping station was also found to be remarkably warm, in part because of heating from the Earth's interior. The researchers came to a conclusion that this impact could likely observed in prior coalfields across Britain.

It is important to ask that is higher rates of warming in these places be instigated by the Earth losing its internal "heat shield"? The notion behind the belief that some regions have a shielding layer below the ground, averting heat from the Earth's interior intensifying to the surface, isn't as odd as it may sound.

After all, the ozone layer present in the atmosphere guards against ultraviolet radiation, but it was only discovered in the mid of 19th century. Surprising new findings about the Earth system emerge continuously from time to time.

In case a related heat-trapping shield present in the Earth's crust, much more should be done to reinforce it. Carbon emissions that are discharged by the industry and energy generation plants could be amassed in the crevices left by extracted fossil fuels, re-insulating the sub-surface and contributing to sluggish the thermal emissions that could be exaggerate global warming.

Scientists are of the opinion that for some certain hope of halting catastrophic climate change have been exists but rests on leaving fossil fuels in the ground.

One major finding in this area revealed that warning is a new urgency. Underground reserves of oil have been in existence for a far longer period of time than humans have exploited them-there is little evidence available about the outcome of emptying them.

2.8. CONVENTIONAL VS. UNCONVENTIONAL OIL

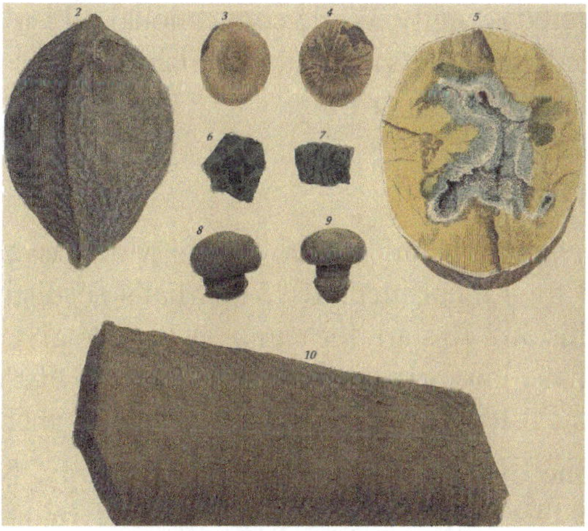

Figure 2.5. A pictorial asset of fossil remains.

Source: Image by Flickr.

Whether a person a new to the oil industry or a seasoned veteran, he or she must likely have heard about the terms "conventional oil" and "unconventional oil," but there are only few who know actual meaning of these terms (Figure 2.5).

This term has, in general, can be a bit misleading, but the variation lies in the method of extraction rather than the oil itself. In the above section, the basic distinction between both has discussed. How both methods can contribute to a productive oilfield is also explained in the following section.

2.8.1. Conventional Oil

It is generally seen that when most people think of oil drilling, they visualize large oil rigs with drills that excerpt oil from hefty pools. This sort of extraction has been in existence since the late nineteenth century, as the oil attained is already running between rock formations underground. The natural pressure from the well is all that required to carry oil to the surface. The term "conventional oil" refers to oil which is extracted by this method.

According to a 2b1 Consulting, when oil is extracted from these oil fields up to its maximum capacity, the well's natural pressure might drop to a certain level where it cannot extract more oil. In this case, gas or water injection can be used. Extraction of oil via these methods is still considered "conventional." After decades of traditional oil extraction, various oil pools have been discovered recently. While conventional oil isn't going that long way, the acknowledgment of its restricted availability led to the advancement of substitute techniques of oil drilling.

2.8.2. Unconventional Oil

The term "unconventional oil" refers to oil which is extracted without using any of the traditional methods. Yet what's so distinctive about the whereabouts of this oil? To start, formations which do give any space for oil flow are said to have low permeability. It is believed that the rocks having permeability below 0.1 millidarcies are deliberated as unconventional.

Both sandstone of coal come under this category. Natural Gas now explains that another imperative distinction concerning how the oil is trapped within these sediments. Extraction of oil by traditional methods is relatively easy because a rig can be drilled straightaway down to the pool.

However, sediments such as sandstone are formed in layered deposits. Instead of gathering in a pool, oil in these layers is present in small amounts. To reach the dispersed oil, the use of horizontal drilling is taken into operation into the layer. One widely recognized and popular is hydraulic fracturing (a.k.a. "fracking"), used to lay cracks or holes in this rock to allow for oil flow.

While the latest technology and techniques open the gate for more unconventional drilling, these techniques are comparatively more cost-intensive than conventional drilling. Some other widely used methods such as fracking have also come under scanner because of their potentially negative impact on the environment.

Despite the terms, both conventional as well as unconventional oil methods are concerned with extraction of oil. While unconventional methods have gained acceptance in the current time frame, conventional oil methods are still used by many in some areas.

It is worth noticing that when both methods are used together, it helps in maximizing the overall efficiency and efficiency of plants. It is important to have full clarification of these terms and related terms in order to gain better understanding about the oil drilling industry as a whole.

2.8.3. Shale Oil and Gas

Shale gas and oil can be defined as natural gas and oil from shale formations. The shale can act as both the reservoir and the source for these unconventional hydrocarbons. Older shale wells were vertical in nature while more topical wells are primarily horizontal and require artificial stimulation, for instance hydraulic fracturing, to produce. Only shale formations have unique characteristics possess the ability to produce gas and oil.

Shale oil is considered as very high-quality crude oil that lies between layers of shale rock, impermeable siltstone or mudstone. Oil companies produce shale oil by shattering the rock formations that contain the layers of oil.

Since the year 2014, U.S. shale oil has witnessed significant role in production of domestic crude oil.

It is important to note that shale oil has included more than one-third of the onshore production of crude oil in the lower 48 states. Because of this, the United States now stands at the position of world's largest crude-oil producer, as per the data revealed by EIA. The dependence on foreign oil imports in the United States has fallen from 9.8 million barrels per day in 2013 to 8.6 million b/d in 2019.

First of all, oil companies drill vertically into the shale formation. After the well is properly drilled, they turn it to 90-degree angle. At this kickoff point, they started drilling horizontally.

This technique opens the opportunity for them to get the access to 10,000 feet of the reservoir rock. They then pump high-pressure bursts of sand, water, and chemicals to fracture the shale in order to release the oil.

The sand makes it possible for fractures to remain open. That allows the oil to seep into the well.

Oil producer plants in Texas are using horizontal drilling technology since 1990s. This method becomes more economical when Brigham Oil and Gas successfully split a single horizontal leg. The company becomes successful in fracturing each leg independently, resulting in getting more return on the investment. This technique triggered the production of the North Dakota field from 7 million barrels in a month to 46 million in a month in 2019.

Because of using this technology, The Bakken Field in North Montana and Dakota became the largest producer of shale oil reserve. The field has layers of dense, oil-bearing rock around two miles underground. In these fields, drillers generally make use multistage fracking in order to create longer cracks. They perforate short segments of the production casing. It plays a very important role for them to retain their focus on bursts of water in targeted spots.

2.9. REFINING: CRUDE OIL

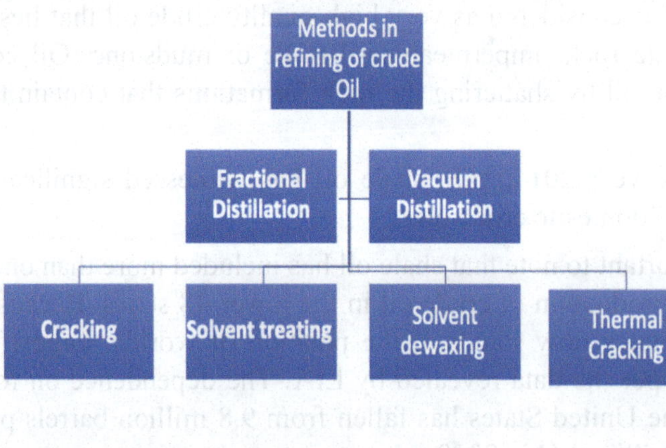

Figure 2.6. Methods in refining of crude oil.

Different hydrocarbons which vary in appearance and composition from one oil field to another together form crude oil. The crude oil is basically a complex mixture. From water to tar-like solids the consistency range of crude oil lies. In case of color crude oil, it lies from clear to black (Figure 2.6).

Oxygen, metals, salts, and nitrogen 1%, 84% carbon, 14% hydrogen, 1–3% sulfur together are the composition of an average crude oil. Based on the predominant proportion of similar hydrocarbon molecules, the crude oils are generally classified as paraffinic, naphthenic, or aromatic.

Further, varying amounts of each type of hydrocarbon molecules are present in the mixed-based crudes. Two or more different crude oils are usually consisted by the refinery crude base stocks.

Fractionation of crude oils into separate hydrocarbons and distillation are the two process which leads to the refining of petroleum. The characteristics of the crude processed are almost directly related to the resultant products.

By changing the size and structure of the hydrocarbon group's molecules through cracking, reforming, and other conversion as discussed below, most distillation products are further converted into more usable products.

To remove undesirable constituents and improved product quality, these converted products are subjected to various treatments. This treatment includes separation processes such as extraction, hydrotreating, and sweetening.

Further, conversion, treatment, blending, and fractionation are incorporated in integrated refineries. Processes such as petrochemical processing might also be included along with the above-mentioned process in integrated refineries.

By distillation in atmospheric and vacuum towers the separation of crude oil into various fractions or straight run cut and this process of separation is the very first step. In order of decreasing volatility into gases, light distillates, middle distillates gas oils, and residues these basically helps to classify the main fraction, or cuts obtained which possess specific boiling point.

2.9.1. Fractional Distillation

By using recovered process heat, the desalted crude feedstock is preheated, at the refinery. In a direct-fired crude charge heater then the feedstock then flows.

At a pressure slightly above atmospheric at a temperature which is ranging from 650 to 700°F (heating crude oil above this temperature and moreover thermal cracking can cause undesirable changes) and this temperature is maintained by the flowing the feedstock into direct-fired crude changer heater and then it is fed into the vertical distillation column just above the bottom.

The heaviest fractions of feedstock flash into vapors it also flows into all. The temperature is reduced by circulating refluxes, and this is due to the hot vapor which rises in the tower.

From bottom, the reduced crude oil (RCO) is taken. Naphtha, Gasoline, Kerosene, Diesel, and uncondensed gasses (these are condensed at lower temperatures) are the various major products which are included at successive higher points on the tower and they are drawn off.

2.9.2. Vacuum Distillation

Reduced pressure is required to prevent thermal cracking, and this is done by further distillation of the residuum or topped crude from the atmospheric tower at higher temperatures. In one or more vacuum distillation towers this process basically takes place.

Further, process such as fractional distillation is resembled by the principles of vacuum distillation. Further to maintain comparable vapor velocities at reduced pressures larger diameter column is used except that fractional distillation and the equipment mentioned are also similar.

Random packing and demister used instead of trays as the internal designs of some vacuum towers are different from atmospheric towers. For the propane de-asphalting, a typical vacuum tower may produce lubricating-oil base stocks, gas oils, and heavy residual. To separate catalytic cracking feedstocks from surplus residuum the vacuum towers are used typically.

2.9.3. Cracking

Subsequent refinery processes change the product mix by altering the molecular structure of the hydrocarbons as the simple distillation of crude oil produces amounts. And further, these products are not consistent with those required by the marketplace.

Cracking is one of the ways of accomplishing this change. More valuable products such as gasoline, fuel oil, and gas oils are obtained by the higher boiling point petroleum fractions, and cracking is a process which also breaks this higher boiling petroleum fractions.

2.9.3.1. Catalytic Cracking

In order to increase the quality and the quantity of lighter, more desirable products and the decrease the amounts of residuals the process of catalytic cracking is used to break the hydrocarbons into simpler molecular.

Further, to convert heavy hydrocarbon feedstocks into lighter fractions which are namely LPG heating oil, kerosene, gasoline, and petrochemical feedstocks this process rearranges the molecular structure of the hydrocarbon compounds.

The pressure is fat much lower pressures of 10–20 psi and the temperature typically ranges from 850–950 degrees. The solid materials which comes in the form of powders, beads, pellets or shaped material called extradites are the catalyst which are used in the refinery cracking units.

These solid materials are namely (zeolite, aluminum hydro silicate, treated bentonite clay, fuller's earth, and bauxite and silica alumina.

The three types of catalytic cracking processes are:

1. Fluid catalytic cracking (FCC);
2. Moving-bed catalytic cracking; and
3. Thermo for catalytic cracking (TCC).

The process which is very flexible is catalytic cracking. To meet changing product demand the operating parameters can be adjusted. The activities such as hydrogenation, dehydrogenation, and isomerization are also included in the catalytic activities in addition to the process of cracking.

The process in which oil is cracked is the most common process which is termed as the FCC. This process is basically done in the presence of finely divide catalyst by the oil vapors this catalyst is maintained in an aerated or fluidized state.

The sections which operate together and acts as an integrated processing which are consisted in the fluid crackers are the catalyst section and a fractionating section. The catalyst circulation unit is the riser forms and along with the standpipe and this is done by the catalyst section which contains the reactor and regenerator. By using air, oil vapors and the steam as the conveying media the fluid catalyst is continuously circulated between the reactor and the regenerator.

A preheated hydrocarbon charge with hot is mixed with the regenerated catalyst in the process of a typical FCC process and this mixing is done as it enters the riser leading to the reactors. By the means of hot catalyst, the charge is combined with a recycle streams within the riser, vaporized, and raised to rector temperature (900–1000°F). The charge is cracked at 10–30 psi when the mixture travels up the riser.

All the cracking takes place in the riser in the more modern FCC units. The reactors serves as a holding vessel for the cyclones as these reactors are no longer functions as reactors. Until and unless the oil vapors are separated from the catalyst in the reactor cyclones till then the cracking continues.

To get rid of coke that collects on the catalyst during the process the spent catalyst is regenerated. Where most of the coke deposits burn off at the bottom where preheated air and spent catalyst are mixed the spent catalyst flows through the catalyst stripper to the regenerator. To optimize the cracking process the fresh catalyst is added and worn out and the catalyst is removed.

2.9.4. Solvent Treating

Solvent treating is one of the mostly used methods that are concerned with refining lubricating oils, in addition to serve as a host of other refinery stocks. Since distillation (fractionation) plays a very important in separation of petroleum products into groups via boiling process-point ranges, impurities may remain a cause of concern. These involve organic compounds comprising nitrogen, sulfur, and oxygen; dissolved metals and inorganic salts; and soluble salts that were existent in the crude feedstock.

The primarily objective of solvent extraction is to enhance the quality of improve finished products by eradicating unsaturated, aromatic hydrocarbons from grease and lubricant stocks. The solvent extraction is mainly interested in separation of naphthene, aromatics, and impurities from the product stream by liquefying them in solvent.

The solvent is then alienated from the product stream by evaporation, heating, or fractionation, and residual trace amounts are consequently distant from the raffinate via steam stripping. The solvent thus regenerated could be again used in the process.

The commonly accepted extraction solvents are furfural, phenol, and N-Methyl Pyrledene (NMP). Some other solvents are not much used such as nitrobenzene, liquid sulfur dioxide, and dichloroethyl ether. The selection of chemical agents and specific processes relied on the environment of the feedstock being treated, the presence of contaminants, and the necessities of finished product.

2.9.5. Solvent Dewaxing

Solvent dewaxing is a process that is primarily used for removal of wax from either residual or distillate base stocks at any phase in the refining process. There are multiple stages and processes exist in solvent dewaxing, but, majority of them have similar general steps, which are discussed as follows:

- Mixing of feedstock with the help of a solvent;
- Precipitating the wax from the resultant mixture via the chilling process; and
- Recovering the solvent from the dewaxed oil and wax for recycling by steam stripping and distillation process.

In majority of the cases, solvents make use of methyl ethyl ketone (MEK) and Propane. Other solvents such as benzene, methyl isobutyl ketone, Toluene, ethylene dichloride, petroleum naphtha, sulfur dioxide and methylene chloride are used sometimes when other options are not available. In addition to this, there is a catalytic process which is used as an alternate to solvent dewaxing.

2.9.6. Thermal Cracking

Thermal cracking processes mainly comprise of coking and include visbreaking that shatter heavy oil molecules by revealing them to high temperatures conditions.

2.9.6.1. Visbreaking

Vacuum residuals and topped crude are heated and thermally cracked (455 to 480°C, 3.5 to 17.6 kg/cm^2 [850 to 900°F, 50 to 250 pounds per square inch gauge (psig)]) in the visbreaker furnace in order to lessen the pour point, or viscosity of the charge.

The cracked products are satiated with gas oil and speeded into a fractionator. The vapor overhead from the fractionator is then alienated into agile distillate products. A heavy weight distillate recuperated from the fractionator liquid can be used for multiple purposes such as a catalytic cracking feed or as a component of fuel oil blending.

2.9.6.2. Coking

Coking is basically referring to a thermal cracking process that is used for converting low value residual fuel oil to higher petroleum coke and value gas oil. Vacuum residuals and thermal tars are split in the Coking process at low pressure and high temperature level.

Products are gas oils, petroleum coke, and lighter petroleum stocks. It is generally seen that delayed coking is the most widely used process in the present time scenario, but fluid coking has higher ability and potential from future perspective.

In the process of delayed coking, heated charge stock is fed into the bottom of a fractionator, where light ends are removed or wrenched from the feed. The stripped feed is then merged with reprocess products from the coke drum and then speedily warmed in the coking heater from temperature range of 480 to 590°C (900 to 1100°F).

For this, steam injection is used in order to govern the residence time in the heater. The vapor-liquid feed separates the heater, transfer to a coke drum where, with governed residence time, temperature (400°C [750°F]), and pressure (1.8 to 2.1 kg/cm^2 [25 to 30 psig]), it is shattered to form vapors and coke. Vapors from the drum re-arrive to the fractionator, where it is possible to recover the thermal cracking products. In the fluid coking process, characterized by Flexicoking, residual oil feeds are instilled into the reactor, where they are shattered thermally, producing a variety of vapor products and a coke.

Vapors then separate from the reactor and are slaked in a scrubber, where there is removal of coke fines. The vapors are then fractionated. Coke from the reactor come in contact of a heater and is devolatilized. The volatiles from the heater are treated for removal of sulfur and fines in order to produce a particulate-free, as well as low-sulfur fuel gas.

The devolatilized coke is scattered from the heater to a gasifier where almost 93% of the reactor coke is gasified at high temperature conditions with air and steam or oxygen. The gaseous products as well as coke from the resultant gasifier are restored to the heater in a way to supply consistent heat for the devolatilization.

These gases then get out of the heater through the sulfur removal and same fines processes because of heater volatility. According to the current knowledge and understanding, it is not clear that what types of emissions are released and where they are released. Air emissions that are released from thermal cracking processes involve coke dust due to decoking operations, fugitive emissions, and ignition gases from the coking process and visbreaking heaters.

Types of emissions from the process heaters are demonstrated below. Fugitive emissions from diverse leaks are substantial because of involvement of high temperatures, and heavily relying on the type of equipment and configuration, general maintenance activities, and operating environments. Fugitive emissions are also explained below. Particulate emissions because of delayed coking operations are also potentially imperative.

These emissions are majorly concerned with removal of coke from the coke drum and subsequent storage and handling operations. It is also observed that the hydrocarbon emissions are also allied with venting and cooling the coke drum before removal of coke. However, comprehensive information for delayed coking emissions is not covered in available literature.

Particulate emission control can be achieved in the operation of decoking by wetting down the coke. Largely, there is no mechanism of hydrocarbon emissions from delayed coking. However, some services are now storing coke drum emissions in an enfolded system and routing them to a refinery flare.

2.10. THE REFINING OF NATURAL GAS

Figure 2.7. Energy development.

Source: Image by Wikipedia.

Refining of the gas is a lengthy process as the gas passes through various processes before it finally reaches our home. The refining of the gas comes ashore from the offshore sub-sea pipeline. The refining process basically comprises of the separation of all the fluids, as well as the hydrocarbon condensates from the oil in order to produce dry natural gas (Figure 2.7).

It is generally seen that most of the gas is CH_4 with traces of other gasses that have to be eradicated, but the stage at which the gas has entered the refinery is discussed. In this section, the process starts after the basic processing has been carried out offshore where condensate and water removal has taken place.

After this, the gas has been transferred to the refinery from the offshore pipeline, but these pipelines tend to clog up with condensates and have to be pigged.

2.10.1. Pigging the Subsea Pipeline

The liquid of condensed build up in these pipes and they have to be pigged every so often. This resulted in conveying a neoprene pig, similar like a bullet down the inside of the pipeline. The pig is a spotless fit and is pushed along by the high pressure of gas behind it as it removes the accretion of condensate from inside the pipe. The pig is propelled from the platform by a pig launcher and is gathered at the refinery by a receiver of pig.

2.10.2. Slug Catcher

The slug catcher refers to pressure vessel which is of horizontal shape and that ensures the removal of any remaining globules of liquid condensates or hydrocarbons before the gas transferred to the next stage. The liquids and condensates are shipped generally by the help of a pipeline to the closest oil refinery for additional fractionalizing.

2.10.3. Acid Gas Removal Unit

This unit primarily comprises of a dryer that help in removal of water, an amine tower for removal of contaminates of CO_2, hydrogen sulfide, water vapor, and oxygen. In addition to this, it also included nitrogen rejection units that help in removal of nitrogen.

2.10.3.1. Dehydrator

The dehydrator is responsible for removal of water moisture content from the gas by injecting the material of glycol liquid into the stream. This operation is undertaken in the glycol tower and as the glycol particles absorb the water they collapse to the base, the dry gas mounting to the topmost for the next process.

The glycol is again generated by boiling of the water that has a lower flash point, allowing the glycol for reuse.

2.10.3.2. Amine Tower

The gas after this transferred to the amine tower where it has the amine has been injected into its stream in order to remove the CO_2 and sulfur contaminates. The gas then conveyed through a nest of filter pipes where it loses velocity and the residue of contaminates tumble to the lowermost sections. The gas when successfully reached through cyclones where centrifugal force knocks out any residual small particles or water vapor.

The hydrogen sulfide, amine solution, and CO_2 can be distant from the lowermost of the tower and sent for treating to produce sulfur that has commercial value. It is a powdery like substance of yellow in color produced in different countries across the world.

When the gas sweetening process is finished, the residue is propelled from the nethermost of the amine tower to the sulfur reclaiming plant. The central of this is the Claus Process that uses thermal and catalytic reaction to reclaim the element.

2.10.4. Nitrogen Rejection Unit

This unit is majorly responsible for removal of nitrogen from the gas by engaging in different repetitive process. Nitrogen and CH_4 condense from gas to liquid at distinct temperatures level. The vapor is cooled to a greater extent; the CH_4 in the vapor then transformed into liquid and is amassed as it drops out from the nitrogen.

The nitrogen is expelled out of the vessel and the CH_4 liquid is then again transformed into a gas subtracted the nitrogen

This plant helps in removal of nitrogen from the gas by engaging in numerous processes.

CH_4 and nitrogen condense from gas to liquid at different temperatures. The vapor is then cooled at a very intense temperature; the CH_4 in the vapor become liquid and is gathered as it drops out from the nitrogen. The nitrogen is then pass out from the topmost of the vessel and the CH_4 which is in liquid form is then returned to a gas subtracted the nitrogen

2.10.5. Methane Separation

There are mainly two reasons for the separation of Natural Gas Liquids (NGLs) from the CH_4 and these are discussed as follows:

- It is a prerequisite for the transfer of natural gas to the networks; and
- The NGLs possess the ability to be refined and used as an alternative for other usable products.

The separation is carried out either by cryogenically or absorption. As a cryogenic method has been earlier examined, now it is important grasp understanding on the absorption method. This is almost related to the method used for dehydration. The only major difference between the two is that it uses an absorption oil to collect the NGLs as compared to dehydration methods, which make use of glycol to amass water vapor.

The gas that carries through a tower where the oil is instilled into the stream and the oil particles soak up the NGLs and fall to the lowermost parting the CH_4 to ensue to the next process.

The oil which is affluent of natural gas liquids (NGLs) now passes to a fractionalization column where the distinctive hydrocarbons are parted out into divisions, condensed, and stowed for used in industrial plants or for commercial sale.

2.10.6. Mercury Removal Process

Removal of mercury from the CH_4 is imperative both from a perception of overall and as a safeguard against loss to the distinctive catalysts used during the process.

This unit usually comprises of pressure vessel, vertical in shape that holds a mutual catalyst of zeolitic sieved, or honeycombed panels soak or saturated with silver commercially known as HgSIL.

The gas then conveyed to the vessel fleeting upwards via sieves, the mercury element being regulated or governed by the dual catalyst.

The gas ensues aloft and out of the top of the vessel and once an odor is enhanced, it becomes prepared for distribution.

2.11. HOW CARBON IS CAPTURED AND STORED?

Figure 2.8. Coal extraction.

Source: Image by Pxfuel.

CCS refers to a range of technologies that plays a very critical role in trapping up to 90% of the CO_2 emissions from industrial sites and power stations. It includes a range of activity such as: collecting, shipping, and then burying the CO_2 in order to make sure that it does not come in contact of atmosphere and altering the climatic conditions (Figure 2.8).

There are basically three main types of techniques: the post-combustion process is concerned with cleaning the power plant's exhaust gas with the help of certain using chemicals. While on the other hand, Pre-combustion CCS activity is undertaken before placing the fuel in the furnace by first transforming coal into a clean-burning gas and then stripping out the CO_2 released by the process. The third method, referred as oxyfuel, burns the coal in an atmosphere at a higher concentration level of pure oxygen, developing an exhaust gas which is almost as pure as CO_2.

Once the CO_2 is being trapped, it is then liquefied, transferred-sometimes for several hundred miles-and buried, either in appropriate geological foundations, disused oil fields, or deep underground saline aquifers. The latter method is widely used in a process known as "enhanced oil recovery," where CO_2 is propelled into an oil field to force out the lasting pockets of oil that otherwise would be challenging to extract.

2.11.1. Trapping Methods

1. ***Post-combustion***: In the post-combustion type of method, CO_2 is alienated from the flue gas of the power station by using the strategy of bubbling the gas through an absorber column enhanced with liquid solvents (for instance ammonia) that preferably excerpt out CO_2. In the widely recognized techniques, once the chemicals present in the absorber column started becoming saturated, a stream of very high temperature at about 120°C is passed through it. This releases the gathered CO_2, which then can be supplied for storage in some other area.

There are many experimental techniques available that can scrub CO_2 from flue gas without even using the two-step process embracing the use of seawater to absorb the gas and then transferred the mixture back to the ocean to store for prolong period of time. But, as of now, these methods are considered less reliable and efficient.

2. ***Oxyfuel***: When the oil, coal, or natural gas is burned in normal air, the portion of CO_2 formed is between 3 to 15% of the waste gases, reliant on the available conditions. Separations of the GHG out after combustion necessitate the high amount of energy so an alternative CCS method is to burn the fossil fuel in an environment of pure oxygen.

In this ecosystem, nearly all the waste gas will be constituted water vapor and CO_2. The latter can be condensed out while the former can be transferred or piped straightforwardly to a storage facility.

In the oxyfuel system, the air that is accumulated into the boiler has to be alienated into liquid oxygen, argon, gaseous nitrogen, and other trace gases and this process can take up to 15% of the power formed at the station.

3. ***Pre-combustion*** A method that is typically applicable at coal-gasification combined cycle power plants. The coal is gasified in order to generate a synthetic gas prepared from hydrogen and carbon monoxide (CO). The former is reacted with water in order to form CO_2, which is captured, and more hydrogen.

The hydrogen then disseminated to a turbine where it can be ignited to generate electricity. In an alternate way, some of this gas can be bled off to feed hydrogen fuel cells for cars.

One major shortcoming of the pre-combustion method is that it cannot be returned back to the traditional coal power plants which make up much of the world's installed base of fossil fuel power. Perhaps, it could be only used in natural gas stations, where first a synthetic gas is formed by reacting CH_4 with steam to form hydrogen and CO_2. But any economic benefit of this method in comparison with post-combustion is yet to be proven.

2.12. CONCLUSION

In the end, it is concluded that there are various types of fossil fuels and each has different kind of extraction process. It depends upon the oil producer which method they want to adopt that is more efficient and effective and that can suit to their needs and wants. While taking various decisions related to formation and extraction of fossil fuels, some of the factors the oil producer should keep in mind as any form of carelessness can affect the climatic conditions to a greater extent.

In addition, there is need to take various cautions while refining of crude oil and natural gas as any negligence can cause significant harm to the persons working in the refining process. CCS process is very crucial to protect the environment as it allows minimizing the amount of carbon that comes in contact of environment.

It is generally seen that fossil fuels are significant contributor to pollution; therefore, the oil producers should take every step towards minimize the impact of these fuels on environment. Moreover, the efforts should be done towards more efficient and effective technology to preserve the fossil fuel for future generation as it is limited in nature.

REFERENCES

1. (2013). Petroleum Refining. [e-book] Available at: https://www3.epa.gov/ttnchie1/ap42/ch05/final/c05s01.pdf (accessed on 4 July 2020).
2. Amadeo, K., (2020). Pros and Cons of Shale Oil Production. [Online] The Balance. Available at: https://www.thebalance.com/what-is-shale-oil-and-how-is-it-produced-3306195 (accessed on 4 July 2020).
3. Api.org, (2017). Refinery Processes. [Online] Available at: https://www.api.org/oil-and-natural-gas/wells-to-consumer/fuels-and-refining/refineries/how-refinery-works/refinery-processes (accessed on 4 July 2020).
4. Bright Hub Engineering, (2014). The Refining of Natural Gas. [Online] Available at: https://www.brighthubengineering.com/marine-engines-machinery/60607-natural-gas-refining/ (accessed on 4 July 2020).
5. Ccsassociation.org, (2020). What is CCS?—the Carbon Capture and Storage Association (CCSA). [Online] Available at: http://www.ccsassociation.org/what-is-ccs/ (accessed on 4 July 2020).
6. Croft system.net, (2014). How Do You Process Natural Gas? [Online] Available at: https://www.croftsystems.net/oil-gas-blog/how-do-you-process-natural-gas (accessed on 4 July 2020).
7. Dghindia.org, (2017). Shale Oil and Gas | Directorate General of Hydrocarbons (DGH). [Online] Available at: http://www.dghindia.org/index.php/page?pageId=37 (accessed on 4 July 2020).
8. Energyeducation.ca, (2016). Fossil Fuel Formation-Energy Education. [Online] Available at: https://energyeducation.ca/encyclopedia/Fossil_fuel_formation (accessed on 4 July 2020).
9. Hindustanpetroleum.com, (2014). Refining Process | Hindustan Petroleum Corporation Limited, India. [online] Available at: https://www.hindustanpetroleum.com/refiningprocess (accessed on 4 July 2020).
10. Jha, A., (2020). Carbon Capture and Storage: The Three Main Technologies. [Online] the Guardian. Available at: https://www.theguardian.com/environment/2008/sep/05/carboncapturestorage.carbonemissions1 (accessed on 4 July 2020).
11. Ji, G., & Zhao, M., (2017). Membrane Separation Technology in Carbon Capture. [Online] ResearchGate. Available at: https://www.researchgate.net/publication/314286570_Membrane_Separation_Technology_in_Carbon_Capture (accessed on 4 July 2020).

12. Kennan, M., (2018). How is Fossil Fuel Converted Into Electricity? [Online] Sciencing. Available at: https://sciencing.com/fossil-fuel-converted-electricity-5170972.html (accessed on 4 July 2020).
13. Keystone Energy Tools, (2015). Conventional Oil vs. Unconventional Oil | Keystone Energy Tools. [online] Available at: https://www.keystoneenergytools.com/conventional-oil-vs-unconventional-oil/ (accessed on 4 July 2020).
14. Metz, B., Davidson, O., De Coninck, H., Loos, M., & Meyer, L., (2020). Carbon Dioxide Capture and Storage Carbon Dioxide Capture and Storage. [e-book] Cambridge University Press. Available at: https://www.ipcc.ch/site/assets/uploads/2018/03/srccs_wholereport-1.pdf (accessed on 4 July 2020).
15. Naturalgas.org, (2014). Processing Natural Gas Naturalgas.Org. [Online] Available at: http://naturalgas.org/naturalgas/processing-ng/ (accessed on 4 July 2020).
16. Norwegianpetroleum.no, (2019). How is Petroleum Formed? Norwegianpetroleum.No. [Online] Available at: https://www.norskpetroleum.no/en/petroleum-resources/petroleum-formation/ (accessed on 4 July 2020).
17. Petroleum.nic.in, (2019). About Refining | Ministry of Petroleum and Natural Gas | Government of India. [Online] Available at: http://petroleum.nic.in/refining/about-refining (accessed on 4 July 2020).
18. Piccirilli, D. I., (2016). Fossil Fuels | EESI. [online] Eesi.org. Available at: https://www.eesi.org/topics/fossil-fuels/description (accessed on 4 July 2020).
19. ScienceDaily, (2020). Fossil Fuel. [Online] Available at: https://www.sciencedaily.com/terms/fossil_fuel.htm (accessed on 4 July 2020).
20. The Conversation, (2011). Fossil Fuel Extraction could be Contributing to Climate Change by Heating Earth from Within. [online] Available at: https://theconversation.com/fossil-fuel-extraction-could-be-contributing-to-climate-change-by-heating-earth-from-within-121331 (accessed on 4 July 2020).

Chapter 3

Catalysis and Hydrogen Production

CONTENTS

The chapter of catalysis and hydrogen production explains the basic significance of the catalyst in any kind of chemical reaction, along with the different tactics or processes which consist production of hydrogen. This chapter explains the definition of the catalyst, along with the working of a catalyst.

This chapter also includes several types of catalysis, such as homogenous catalysis and heterogeneous catalysis. This chapter explains the basic significance of the necessity of hydrogen in modern times. This chapter provides highlights on how to produce hydrogen and its production process technologies.

This chapter also mentions about producing clean hydrogen, its significance, and its process. This chapter also consisting of the production of hydrogen with the help of fossil fuels, along with its several numbers of various processes such as steam reforming, partial oxidation, auto thermal reforming, water gas shift, preferential oxidation, and methanation. This chapter also explains the importance of the storage of hydrogen.

3.1. INTRODUCTION

The other field of the chemistry which is deeply mingles across all the area of sustainable energy is catalysis. The process of catalysis allows more environmentally friendly production tactics and methodologies and the cuts the cost of the manufacturing by enormously decreasing the amount of the energy in order to carry out the chemical transformations.

Moreover, the enhancements and developments in the catalysis have permitted the synthesis of evermore complicated materials with the greater selectivity. With respect to the energy arena, catalysts are the basic needs in order to refine the petroleum, the operation of the fuel cells, the production of the hydrogen, as well as the synthesis of the most polymers.

This chapter will provide a highlight of what is catalyst, how it works. In addition to this, the chapter will specifically focus on the production of the hydrogen gas, where catalysts play the important role in the production of hydrogen.

3.2. WHAT IS CATALYSIS?

In the field of chemistry, catalysis is the change or alteration of the rate of a chemical reaction, generally acceleration, with the help of the addition of a substance which is not being consumed through the course of the reaction.

The rates of the chemical reactions, which is, the velocity or speed at which the chemical reaction takes place, relies on the several numbers of factors such as the chemical nature of the species, which is being reacted, and the external circumstances to which they are exposed.

A specific kind of phenomenon is linked with the rates of chemical reactions that is of great theoretical as well as practical interest is catalysis, the acceleration of the rate of the chemical reaction to take place with the help of the substances that are not being consumed through the course of the reaction, themselves, these sorts of substances are known as catalysts.

The study of the process of the catalysis is of the interest theoretically, and the reason behind that is that, what is discloses about the fundamental nature of the chemical reactions. In the context of practice, the study of the catalyst is very necessary, and the reason behind that is, several numbers of various industrial processes relies on the catalysts for their success. Basically, the peculiar phenomenon of life would hardly be possible without the help of the biological catalysts, which are known as enzymes.

With respect to the catalyzed reaction, typically, the catalyst enters into the chemical combinations with the reactants but is regenerated, in the end, in such a way that the amount of catalyst remains unchanged.

As all know, that the catalyst is not consumed, each, and every molecule of the catalyst can induce the transformation of several numbers of the molecules of the reactants.With respect to an active catalyst, the numbers of molecules that are transformed per minute with the help of one molecule of the catalyst can be as large as the several million.

Where a given substance or a combination of the substances go through two simultaneous reactions or more than two simultaneous reactions that produces or develops different products, the allocation of the products can be impacted with the help of the use of the catalyst that selectively increases the rate of one reaction as it is compared to the other reactions. With the help of opting suitable catalyst, a specific kind of reaction can be made in order to take place to the amount of practically excluding the other. Several numbers of various significant applications of the catalysis are based on the selectivity of this type (Figure 3.1).

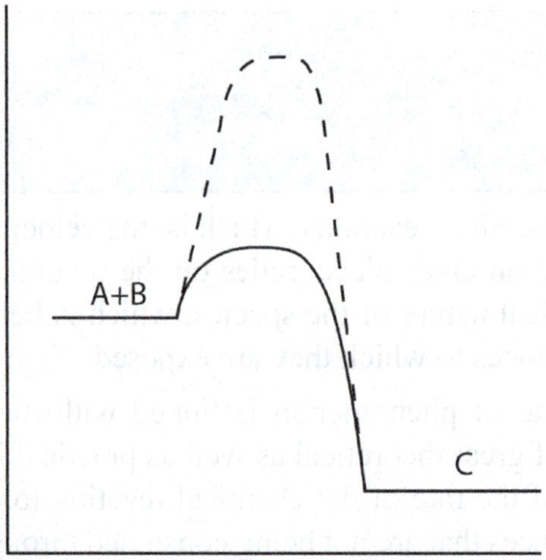

Figure 3.1. An illustration of exothermic reaction with catalyst.

Source: Image by Wikimedia commons.

As, a reverse chemical reaction can proceed with the help of the reversal of the steps that made up the mechanism of the forward reaction, the catalyst for a given reaction increases the rate of the chemical reaction in both directions in equal amount (Figure 3.2).

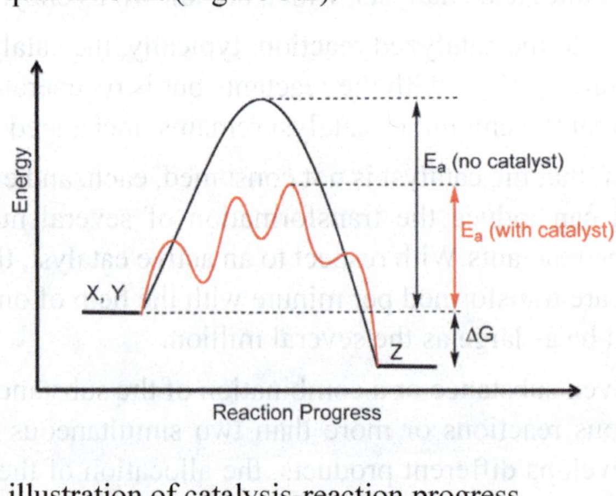

Figure 3.2. An illustration of catalysis-reaction progress.

Source: Image by Wikimedia commons.

In this way, a catalyst does not impact the position of the equilibrium of the chemical reaction. The catalyst only impacts the speed at which the equilibrium is accomplished. Apparent exceptions to this generalization are those reactions in which one of the products is also work as a catalyst in the reaction. These types of chemical reactions are known as autocatalytic.

Several cases are also known in which the addition of a foreign substance, which is known as an inhibitor. The inhibitor reduces the acceleration or rate of the chemical reaction. This phenomenon is known's as inhibition or retardation, and every so often, it is also known as negative catalysis.

Concentrations of the inhibitor may in some of the cases be much lower when it is compared to those of the reactants. Inhibition can result from, the following conditions:

1. A decrement in the concentration of one of the reactants due to the complicated formation between the reactant and the inhibitor.
2. A decrement or reduction in the concentration of an active catalyst ("poisoning" of the catalyst) because of the complicated formation between the catalyst and the inhibitor, or
3. A termination of the chain reaction and this takes place due to the destruction of the chain carriers with the help of the inhibitors.

3.3. CLASSIFICATION OF CATALYSIS

Generally, the catalysts can be categorized as per their physical state, the chemical nature of the catalysis, or the nature of the reactions that they catalyze.Catalysts can be present in the form of gases, liquids, or solids. In the context of homogenous catalysis, the catalyst is molecularly dispersed in the same phase (generally, it is gaseous phase or liquid phase) as the reactants.With the respect to the heterogenous catalysis, the reactants and the catalyst are present in different phase, and they are separated with the help of a phase boundary. Most generally, the heterogenous catalysts are solids, and the reactants are present in the form of gas or liquid.

3.3.1. Homogeneous Catalysis

When the catalysts along with the reacting substance are present together in a same or single state of matter, typically they are present in the form of gas or a liquid. It is normal and expected to categorize the reaction as the cases of the heterogenous catalysis.

Oxides of the nitrogen serve as the catalysts in order to oxidize the sulfur dioxide inside the lead chamber process in order to generate the sulfuric acid. As an example of homogenous catalysis the catalyst along with the reactants and both of them are present in the form of gases.

Traces of vapors of water catalyze some gas reactions. For instance, the interaction of the carbon monoxide (CO) and the oxygen, which as an outcome, only shows in the dry settings. Sulfuric acid is used in the form of catalyst in order to produce diethyl ether with the help of the ethyl alcohol, which is an example of homogenous catalysis in the liquid form (when the products, water, and ether, are frequently removed with the help of the distillation).

With the help of this method, the significant amount of alcohol can be transformed into the ether with the help of a single charge of sulfuric acid. The inversion of sugar cane and hydrolysis of the esters with the help of the acid solutions are also the examples of the homogenous catalysis that takes place in the liquid phase (Figure 3.3).

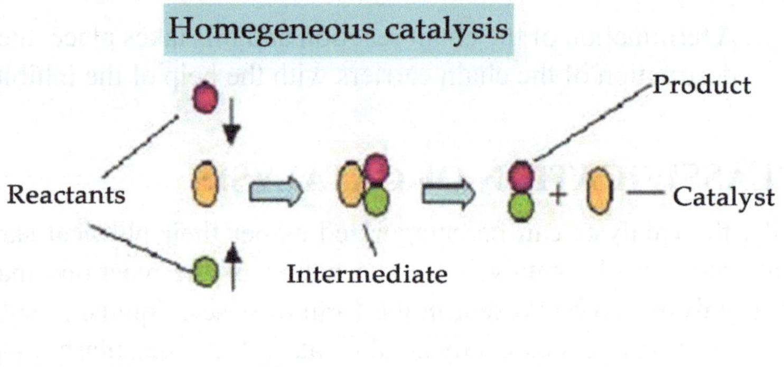

Figure 3.3. An illustration of homogeneous catalysis.

Source: Image by Wikimedia commons.

The oxidation of the sodium sulfite solution with the help of the dissolved oxygen is more and more increased by minute traces of the copper ions in the homogenous liquid system. This system is of special interest from the time it has been shown that the process is a chain reaction.

With respect to this case, several thousands of the molecules of the sodium sulfite can be oxidized into the form of sulfate if the initial activation process is produced with the help of the absorption of a limited number of quanta (which is discrete energy measures) of the light.

The best instance of a chain reaction, which is initiated with the help of the light, is the photo combination of the hydrogen (H_2). With respect to this context, as many as one million molecules of the hydrogen chloride can be developed with the help of the absorption of a single light quantum (which is denoted as h_v). The sequence of the chemical reaction is mentioned below:

$$Cl_2 + h_v = 2\ Cl \quad (1)$$

$$Cl + H_2 = HCl + H \quad (2)$$

$$H + Cl_2 = HCl + Cl \quad (3)$$

With the chemical reaction (2) and chemical reaction (3) over and over again, it is of interest to note that these types of chain reactions can be slowed down with the help of the presence of the negative catalysts, which is generally known as the term inhibitors.

Inhibitors are the substances that slow down the rate or the acceleration of the chemical reaction by shortening the reaction chain. More typically, with the help of entering into a non-chain chemical reaction with one of the chemical compounds that maintain the chain.

An extensive variety of the substances, which is consisting of the alcohols, sugars, and phenols, that have been found in order to play the role of the inhibitors or negative catalysts of the sulfite solutions.

A generalized treatment of the heterogeneous catalyst with the help of the acids and bases was given by the Danish physical chemist J.N. Brønsted in the time period of mid-1920s. This treatment is given on the basis of his concept of acids and bases. As per the treatment of the Brønsted, an acid is a molecule that can furnish a proton, and a base is a molecule that takes up a proton.

As per this assumption, the range of the acids that are consisting of these types of several various materials as bisulfate ion, HSO_4^-, acetic acid, CH_3COOH; water, H_2O; hydronium ion, H_3O^+; and ammonium ion, NH_4^+. The corresponding bases are sulfate ion, SO_4^{2-}; acetate ion, CH_3COO^-; hydroxide ion, OH^-; water, H_2O; and ammonia, NH_3 (these several substances accept the protons in order to produce the listed acids).

Brønsted studied a number of acid-and base-catalyzed reactions that are consisting of:

1. The acid-catalyzed hydrolysis of an ester, ethyl orthoacetate;
2. The basic catalysis of nitro amide decomposition ($H_2N + NO_2 \rightarrow H_2O + N_2O$); and
3. The acid-base catalysis of the conversion (mutarotation) of

glucose to a closely related form. In each case, he observed a direct relationship between the velocity of the catalyzed reaction and the concentration of the catalyzing substance.

Parts that are based on the idea of the Brønsted, a general scheme for a change of a substance A into the other substance B catalyzed with the help of the material C, can be formulated as mentioned below:Based in part on the ideas of Brønsted, a general scheme for a change of a substance A to another substance B catalyzed by a material C can be formulated thus:

$$(1) \qquad A + C \underset{k_2}{\overset{k_1}{\rightleftharpoons}} Z$$

$$(2) \qquad Z \xrightarrow{k_3} B + C$$

The designation Z refers to an intermediate stage. That intermediate stage is formed with a velocity, which is indicated with the help of K_1, Z can disappear either in order to reform A + C, with a speed of K_2, or it can decompose with the help of path 2, with the velocity K_3, in order to yield the product B and regenerate the catalyst C.

If K_3 is much greater as compared to the K_2, the intermediate Z is used up almost as quickly as it formed. The product will then be formed at the rate which is governed with the help of expression k_1 [A] [C]. With respect to the expression, the square brackets show the concentration of both, the concentration of the reactant as well as the concentration of the catalyst.

If K_2 is much larger, when it is compared with the K_3, never the less, the velocity-determining process is the decomposition of Z, the speed or the rate of the formation of the product being represented with the help of the expression k_3 [Z]. With respect to this expression, [Z] is the concentration of the intermediate. Examples of both types of change have been studied.

With respect to the certain cases, two catalysts or more than two catalysts are present at the same point of time produce effects which is greater than either would produce alone. It is then customary to speak of the promoter action. In this way, the ions of the iron are the solution fortifies the action of copper ions in catalyzing a reaction between the hydrogen peroxide and iodine. It is expected that each and every catalyst trigger only one of the reactants.

As per the modern interval of times, the most important example of catalyzes are found in the petrochemical industry. The oxo reaction is one of these types of process. With respect to the oxo reaction, the CO and the hydrogen are added to olefins (which is an unsaturated hydrocarbons) at near about hundred and 50°C (near about 300°F) and two hundred atmospheric pressure in order to produce aldehydes and alcohols, as well as the organic compounds that are containing oxygen.

A cobalt carbonyl catalyst $Co_2(CO)_8$ is used. This hydrocarbon-soluble catalyst is supposed to trigger the hydrogen with the help of formation of the H-Co $(CO)_4$, after then it reacts with the olefins. This reaction has led to several numbers of various studies of organometallic chemistry.

The cations of the copper, silver, along with the mercury (which are positively charged ions) and permanganate anions (which are negatively charged ions) also are known to act as homogenous catalysts with respect to the hydrogen activation.

Palladium chloride is used commercially in the catalytic oxidation of the ethylene into the form of acetaldehyde in the presence of the cupric chloride.

The palladium is presumed to be repeatedly transformed from the salt to the free metal, the function of the cupric chloride being to take participation in the re-information of the palladium salt from the metal.

Phosphoric, sulfuric, sulfonic, and hydrobromic acids play a very significant role of agents in the industrial process of the isomerization, polymerization, hydration, and dehydration, as well as in the classic esterification reactions.

Free radicals (that are molecular fragments bearing unpaired electrons) that are produced with the help of the decomposition of peroxides or metal alkyls also initiate homogeneous catalytic processes.

3.3.2. Heterogeneous Catalysis

Several numbers of catalytic processes are known in which the catalyst along with the reactants are not present in the same phase or in same form of matter, that is the state of matter. These types of chemical reactions are known as heterogeneous catalytic reactions. They are consisting of reactions between the gases or liquids or both at the surface of a solid catalyst.

As the surface is the place at which the reactions take place, typically, it is prepared in such a way that produces large surface areas per unit of catalyst. These areas are finely divided metals, metal gauzes, metals that are combines with the supporting matrices, and metallic films have all been used in the modern heterogeneous catalysis. The metals themselves are used, or they are transformed into the halides, oxides, and sulfides.

With respect to the solid catalysts, at least one of the reactants is chemisorbed (which is hybrid term for chemically absorbed) with the help of the catalyst. A catalyst is chosen that releases the products formed as fast as possible, or else the products remain on the catalyst surface and act as poisons with respect to the process.

Chemisorption can take place over an extensive range of temperature and the most efficient temperature for the adsorption relies on the nature of the catalyst. In this way, the hydrogen is chemisorbed very fast with the help of several numbers of various metals even at the liquid air temperature (which is below–180°C or–290°F).

With a series of hydrogenation-dehydrogenation catalysts, for example, zinc oxide-chromic oxide ($ZnO–Cr_2O_3$), which is basically the process of chemisorption of hydrogen, that every so often takes place above the room temperature. Nitrogen is fastly chemisorbed on the synthetic ammonia-iron catalyst in the region above 400°C (or 750°F).

It has been shown that the films of iron chemisorb the nitrogen even at liquid air temperature, with the additional chemisorption which is found above the room temperature. It follows from these types of considerations that never the less physical adsorptions, which parallel the ease of the liquefaction of the absorbed substance, that take place instantly.

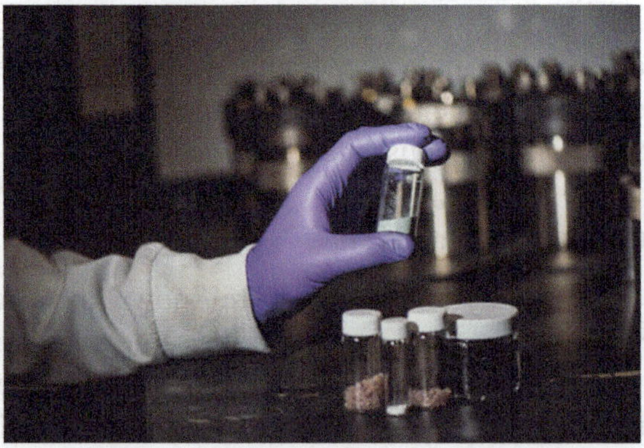

Figure 3.4. An illustration of heterogeneous catalysis.

Source: Image by Flickr.

Chemisorption which is consisting of the making as well as breaking of the chemical bonds, every so often needs activation energies (the energy which is required in order to start the reactions) as do uncatalyzed the chemical processes (Figure 3.4).

In order to be effective catalytically, a process must involve the energies of activation for all the steps involved that, at their maxima, are less when it is compared to those, that are needed for the uncatalyzed chemical reaction.

There are two competing proposals have been made which are having the concern about the mechanism of the catalytic reactions at surfaces, and it has not been possible to opt between them. In the beginning, Irving Langmuir, who was an American physical chemist, proposed chemisorption of both reacting species at the surface, which is followed with the help of interaction between the adjacent species as well as the evaporation of the products.

Apart from that, an alternative proposal involves the interaction between an imposing molecule and species that are already absorbed on the surface. Following developments have suggested the several numbers of different modes of attachment of the adsorbed and adsorbing species.

The most important development in the science of the surface catalysis was the development of an approach or a tactic, which provide assistance in order to find out the surface area of the catalysts (along with the other materials) with the help of measuring the multimolecular adsorption of nitrogen at liquid nitrogen temperatures or the adsorption of the other gases that are close to their boiling points.

It then became possible in order to evaluate or calculate a quantity, known as $V_{m.}$ This quantity depicts the volume of the gas, which is required, in order to develop a monolayer on the accessible surface. In addition to this, the area of the surface can be finding out with the help of the known dimensions of the adsorbed molecules.

It has also been found possible in order to titrate (or to measure quantitatively) the area of the surface with the help of the chemisorption of the gases. As the reactions that are heterogeneously catalyzed take place on the surface of the catalyst, the speed or the rate of these types of chemical reactions are proportional to the accessible surface area of the catalyst.

In this way, the catalysts are generally highly porous solids with the total surface areas as high as several hundred square meters per gram.

When the measurements of the surface areas become possible to take, it was seen at once that several numbers of various components are present in the minor quantities in the main catalyst material; these minor qualities are known as the promoters. The promoters could act with the help of extending the efficient surface areas of the catalyst.

Never the less, it was also known that a promoter can produce an increment in the quality of the surface for that particular given reaction.

3.4. HYDROGEN PRODUCTION

All across the globe, hydrogen is the most abundant chemical element. Hydrogen is found in the sun, also, in the other stars and the gas planets that are present in our solar system. It found naturally on the Earth, but not in large enough quantities to be produced cost-competitively. In this way, it requires to be separated from the other elements.

Talking about the earth, hydrogen is generally found in the compounds along with the other elements. Out of which the most common element is carbon, with the help of which it produces methane (CH_4), and oxygen (O_2). How is pure hydrogen made? In order to obtain pure hydrogen so that it can be used commercially, hydrogen must be separated from the chemical element to which the hydrogen is bound with (Figure 3.5).

Figure 3.5. Basic significance of hydrogen molecule.

Source: Image by Flickr.

3.4.1. Hydrogen Production Process Technologies

In the present interval of time, near about 95% of the hydrogen is produced from the wood or with the help of the fossil fuels, for example, natural gas and oil. There are three kinds of processes that are used in order to produce the hydrogen, that are currently being used:

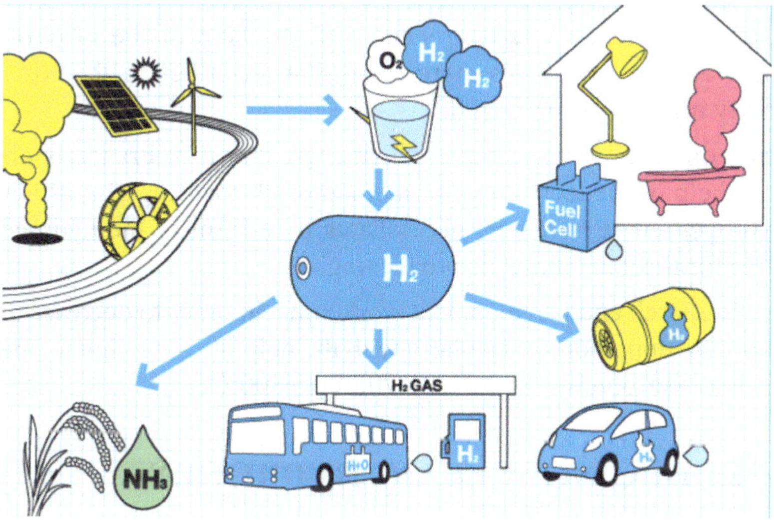

Figure 3.6. The application of hydrogen in several numbers of various sectors.

Source: Image by Wikimedia commons.

Out of the which, the most common process of producing the hydrogen is natural gas reforming, every so often is known as the steam CH_4 reforming, as it utilizes steam which is of high temperature. When it is exposed to the steam and heat, the atoms of the carbon (C) separated from CH_4 (Figure 3.6).

After the two consecutive reactions, they reform separately in order to produce hydrogen (H_2) and the carbon dioxide (CO_2). In this way, this operation necessitates the use of the natural gas.

Apart from this, the second process is known as charcoal gasification. Charcoal is mainly consisting of carbon and water. Charcoal is burned inside a reactor at a very high temperature of between, near about 1200°C and 1500°C. After that, the charcoal releases gas that separates and reforms in order to produce the hydrogen (H_2) along with the CO.

The production of the hydrogen can also take place with the help of the electric current, through the electrolysis of the water. An electric current is used in order to separate the water (H_2O) into the oxygen (O_2) and hydrogen (H_2). This method is not as cost-efficient, when it is compared with one of the fossil fuels.The production of hydrogen with the help of steam CH_4 reforming costs nearly about €1.5 per kilogram at the plant gate (apart from the distribution cost). This cost is near about three folds, when compared with hydrogen production with the help of fossil fuels.

The production of the hydrogen, with the help of the electrolysis is presently around four times more expensive, even before the cost of the electricity required is factored in.

Only one% of the hydrogen which is produced in France is obtained with the help of the technology or method of electrolysis. But as new uses for hydrogen energy emerge, necessitates pure form of the hydrogen, the horizons for this technology is broadening.

Research and Development work aims, in order to decrease the cost of the production, more particularly with the help of using high temperature (or steam) electrolysis at near about 700°C to 800°C.

3.5. PRODUCING "CLEAN" HYDROGEN

In order to make sure that this new energy carrier is "green" or environmentally friendly. In simple words, it releases little amount of greenhouse gas (GHG), for that, the production of hydrogen has to be as clean as possible.

Apart from this, gasification is another option, in order to produce the hydrogen. And the reason behind this is that, it covers the whole solid biomass pathway. Several types of various organic matters can be burned in order to produce biogas.

Never the less, wood (in the form of charcoal) is the main feedstock used, plant waste, for example as straw is also appropriate. And the reason behind that is biomass source can be replanted as the carbon footprint is low.

Electrolysis also generates clean hydrogen when "green" electricity is being used. But, in order to find out the solution with respect to the issue of cost-competitiveness, huge amount of cost-efficient electricity is required, all across the year.

With respect to Germany, there are several various prototypes that are under study, in order to use intermittent production spikes from wind as well as solar energy. But, with respect to present interval of time, the cost of the electrolysis is still very high as compared to the other process of producing hydrogen (Figure 3.7).

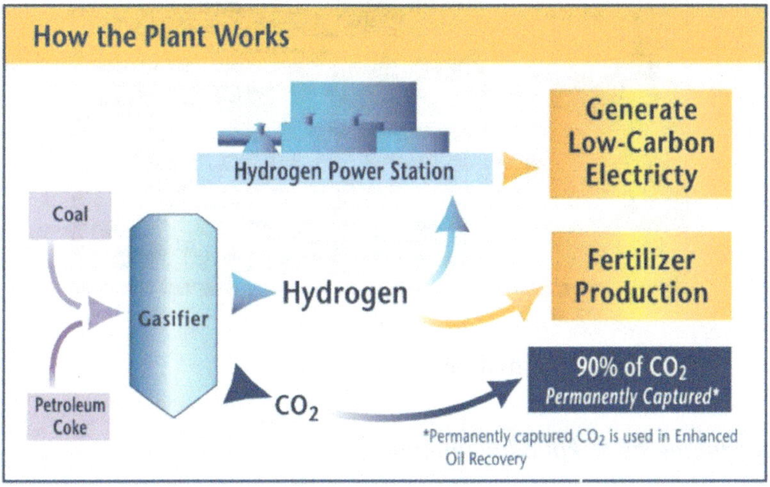

Figure 3.7. An illustration of hydrogen energy process.

Source: Image by Wikimedia commons.

3.5.1. Other Hydrogen Production Processes

Several modified microorganisms can also be used, in order to produce the hydrogen, when they are exposed to the solar radiations or sunlight. This process is known as biological hydrogen production with the help of the photosynthetic microorganisms.

With respect to the photo electrolysis, a photo electrochemical cell, which is placed under the water, uses the solar radiations or sunlight in order to split the water into the form of hydrogen and bubbles of oxygen.

Talking about the thermo chemical cycle, water is being heated to near about 800°C to a 1000°C, with the molecule splitting, in order to release hydrogen. The main limitation of this method is that it necessitates nuclear power in order to heat the water.

In addition to this, the cost of this process of producing hydrogen is very high, along with the cost factor; the production of hydrogen also relies on the stocks of the uranium (Figure 3.8).

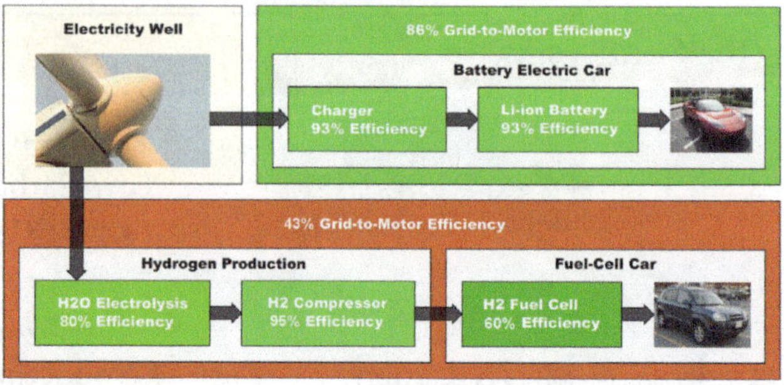

Figure 3.8. Battery EV vs. hydrogen EV.

Source: Image by Wikimedia commons.

3.6. HYDROGEN FROM FOSSIL FUELS

The production of hydrogen can be done with the help of the fossil fuels. The technology of processing fossil fuels to convert them into hydrogen is consisting of materials that are derived from the fossil fuels.

Examples are gasoline, hydrocarbons, methanol, or ethanol, into a hydrogen-rich gas stream. In the present interval of time, fuel processing of CH_4 (which is a natural gas) is the most common method in order to produce the hydrogen commercially.

Most of the fossil fuels are consisting of a specific amount of sulfur, the removal of which is a significant job, with respect to the planning of economy which is based on the production of hydrogen.

The production of the hydrogen gas can be taking place with the help of hydrocarbon fuels through three basic technologies:

(i) Steam reforming (SR);
(ii) Partial oxidation (POX); and
(iii) Auto thermal reforming (ATR).

These three various technologies produce a great deal of CO. In this way, in a following step, one chemical reactor or more than one chemical reactor are used in order to largely transform the CO into the CO_2 with the help of the water-gas shift (WGS) and preferential oxidation (Pr-Ox) or methanation reactions.

3.6.1. Steam Reforming

At the present interval of time, the SR is one of the most extensive and at the same time, most cost-effective process for the production of hydrogen. Its advantage increases from the high efficacy of its operation and the low operational and production costs.

The most frequently used raw materials are natural gas and lighter hydrocarbons, methanol, along with other oxygenated hydrocarbons. The networks of reforming reactions of the hydrocarbons and the methanol which is used as feedstock is the following:

$$C_mH_m + m\ H_2O\ (g) = m\ CO + (m + 0.5\ n)\ H_2 \qquad (1)$$

$$C_mH_m + 2m\ H_2O\ (g) = m\ CO_2 + (2\ m + 0.5\ n)\ H_2 \qquad (2)$$

$$CO + H_2O\ (g) = CO_2 + H_2 \qquad (3)$$

$$CH_3OH + H_2O\ (g) = CO_2 + 3H_2 \qquad (4)$$

The entire process is consisting of two stages. In the very first stage, the hydrocarbon raw material is mixed with the steam and then it is fed in a tabular catalytic reactor. Through the course of this process, syngas (which is the mixture of the hydrogen and CO gas) is generated with lower content in CO_2 (as mentioned in the chemical equation number (1) and (2)).

The desired temperature of the chemical reaction is accomplished with the help of the addition of oxygen on air for combusting part of the raw material (which is heating gas) inside the reactor. In the context of the second stage, the cooled product gas is then fed into the CO catalytic converter, where the CO is then transformed to a large extent with the help of steam into the CO_2 and hydrogen.

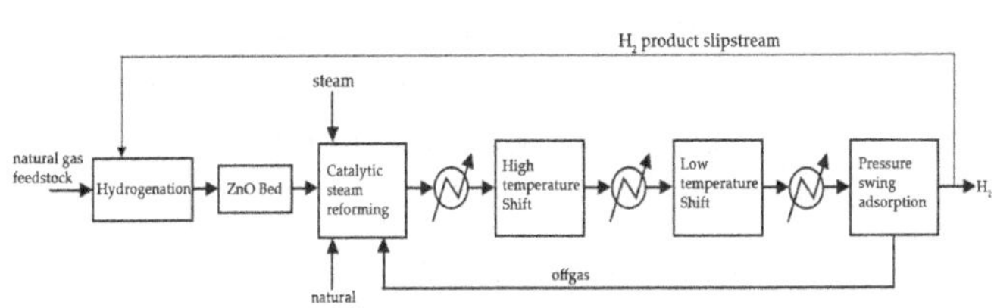

Figure 3.9. An illustration of steam reforming.

Source: Image by Wikimedia commons.

The SR catalytic process necessitates a raw material which is free of compounds that are consisting with sulfur, in order to avoid the deactivation of the catalyst that is being used in the chemical reaction (Figure 3.9).

The process of SR necessitates modest temperature, for instance, hundred, and 80°C for methanol and oxygenated hydrocarbons and over 500°C with respect to the most conventional hydrocarbons.

The catalysts that are being used can be split in to two different types. Out of which, the first and foremost is metal (generally nickel) and precious metals from Group VIII elements (which is generally Platinum and Rhodium). Because of the severe mass as well as limitations of heat transfer, the traditional conventional steam reformer are limited with the help of the efficient factor of pelletized catalysts, which is generally less than 5%.

In this way, the kinetics is hardly ever the limiting factor with respect to the traditional steam reformer reactors, and in this way, they are cost effective nickel catalysts are used commercially.

3.6.2. Partial Oxidation

Partial oxidation (POX) and catalytic partial oxidation (CPOX) of hydrocarbons have been proposed in the production of the hydrogen with respect to the automobile fuel cells along with some other commercial applications.

The gasified raw material could be in the form of CH_4 and biogas. But mainly heavy oil fractions (for example, vacuum remnants, heating oil), whose further treatment and utilization is a very difficult task to perform.

Partial oxidation is a noncatalytic process. The noncatalytic process is said to be a process, in which the raw material is gasified in the presence of oxygen (as shown in the chemical reactions (5) and chemical reaction (6) and possibly steam (7), ATR at temperatures in the 1300°C to 1500°C and pressures in the range of three to eight megapascals. In comparison with the SR (H_2 : CO = 3 : 1), more carbon monoxide CO is produced (H_2 : CO = 1 : 1 or 2 : 1).

In this way, the process is complemented with the help of the conversion of CO with steam into H_2 and CO_2. This reaction contributes to the maintenance of equilibrium between the individual reaction products:

$$CH_4 + O_2 = CO + 2H_2 \quad (5)$$

$$CH_4 + 2O_2 = CO\ 2 + 2H_2\ O \quad (6)$$

$CH_4 + H_2O\ (g) = CO + 3H_2$ (7)

The gaseous mixture is formed with the help of the partial oxidation, which is consisting of CO, CO_2, water (H_2O), hydrogen gas (H_2), methane (CH_4), hydrogen sulfide (H_2S), along with the carbon oxysulfide (COS). A part of the gas is burned to provide enough heat for the endothermic processes.

The soot produced with the help of the decomposition of acetylene as an intermediate product is an undesired product. The amount relies on the quantity of hydrogen by carbon (H:C) in the initial raw fuel material. There has been, in this way, like with the SR, an attempt in order to change to raw materials which are having a higher ratio of carbon by hydrogen (H:C), for instance, to natural gas.

On the other hand, the operation of the reactor is comparatively cost-effective, when it is compared with the SR, the following or succeeding conversion makes this technology more expensive. As the process does not necessitate the application of a catalyst, it is not essential to eliminate or take away the sulfurous elements from the natural gas, which, as an outcome, decreases the effectiveness or efficacy of the catalyst.

The sulfurous compounds are consisting of the gasified raw material and they are transformed into hydrogen sulfide (near about 95%) and COS (near about 5%).

3.6.3. Auto Thermal Reforming

As it has been mentioned in the previous topics as well, in the ATR steam is added in the CPOX process. ATR is an integration of both SR (which is an endothermic process in nature) as well as POX (which is an exothermic process in nature) reactions. ATR has the advantages of not necessitating the external heat and being simpler and cost-effective when it is compared with the SR of CH_4.

The range of operation of a fuel, along with the selection of conditions of operation of the reformer, relies on the particular target. A main target is the high amount of production of hydrogen with less content of CO.

For maximum efficiency of hydrogen, as well as the less content of CO are possible in the SR. Never the less, SR is an endothermic process in nature and therefore energy demanding. This energy has to be transferred into the system from the outside.

Apart from this, the other important or valuable advantage of ATR as compared to the SR process is that, the ATR can be shut down and started very rapidly, through the course of producing a larger amount of hydrogen when it is compared with the process of POX alone.

In addition to this, there are some expectations that this process will become attractive for the "Gas to Liquid" fuel industry, and the reason behind this the favorable gas composition for the Fischer-Tropsch synthesis, relative compactness of ATR, lower need for financial inputs, and the potential for economies of scale.

In the context of CH_4 reforming, the thermal efficiency is comparable to that of POX (which is near about 60% to 75%) and slightly less as compared to that of SR. Gasoline along with the other higher hydrocarbons can be transformed into the hydrogen on board for use in automobiles with the help of the auto thermal process, through the use of appropriate catalysts.

3.6.4. Water-Gas Shift, Preferential Oxidation, and Methanation

The reforming process generates a product gas mixture with the considerable amount of CO, every so often 5 vol% or more (near about 10 vol%). In order to elevate the amount of the hydrogen, the product gas is passed across a WGS reactor in order to reduce the content of the CO, on the other hand, at the same point of time increasing the content of the hydrogen. Generally, a high temperature is required in order to favor the fast kinetics (Figure 3.10).

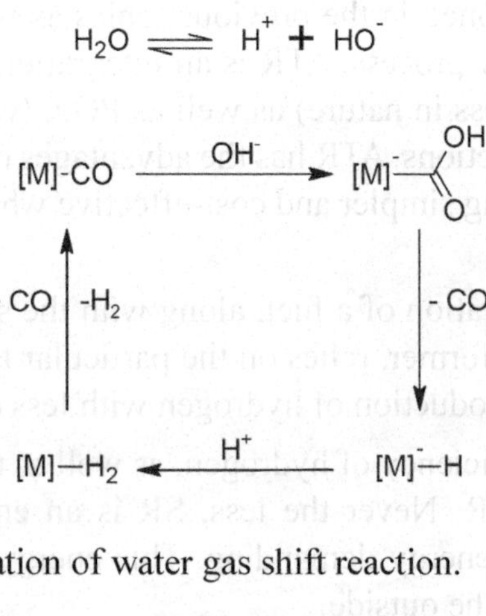

Figure 3.10. An illustration of water gas shift reaction.

Source: Image by Wikimedia commons.

Never the less, consequences are in the form of high equilibrium CO selectivity and reduced production of the hydrogen. In this way, the decrement in the content of CO of syngas is accomplished in a two-step process, which is consisting of a high-temperature WGS reaction and a low-temperature WGS reaction. This reaction is known as high-temperature shift (HTS) and low-temperature shift (LTS) processes, respectively.

With respect to the first step, the reaction is carried out in temperature range of near about three hundred and 10°C to 450°C with the use of Fe_3O_4 / Cr_2O_3 catalyst, the content of the CO is decreased from 10 to 3 vol%.

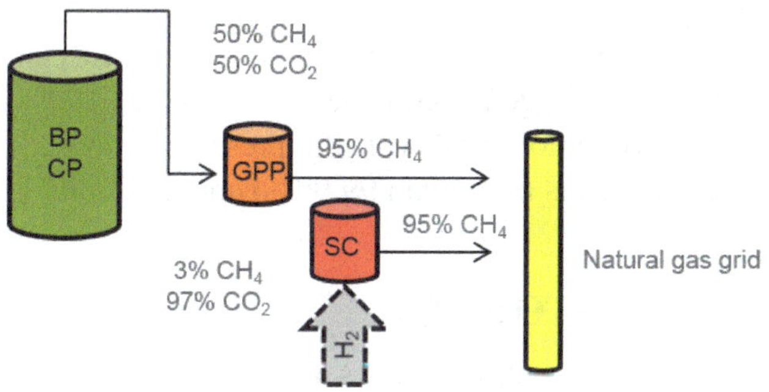

Figure 3.11. An illustration of biological methanation.

Source: Image by Wikimedia commons.

the second step, the reaction is carried out in temperature range of near about 180°C to 250°C, the content of the CO is further decreased to the low level of 500 ppm with the help of the Cu /ZnO/Al_2O_3 catalysts (Figure 3.11).

In order to further decrease the content of the CO in the product gas, a preferential oxidation (Pr Ox) reactor or a CO selective methanation reactor is making into the use. Every so often, the term selective oxidation is used in place of Pr Ox.

Selective oxidation refers to the reduction of the CO within a fuel cell, generally a proton exchange membrane (PEM) fuel cell, never the less, Pr Ox takes place inside a reactor external to the fuel cell. The Pr Ox as well as methanation reactors has their own advantages and challenges, on the same time.

The Pr Ox reactor elevates the complicatedness of the system. And the reason behind this is the precise amount of air must be added to the system. Never the less, these reactors are small in size, and if extra amount of air is introduced, some quantity of the hydrogen is burned.

Methanation reactors are simpler in that no air is needed. Never the less, with respect to each and every CO reacted three molecules of hydrogen (H_2) are consumed. Also, CO reacts with the hydrogen, and careful regulation of the conditions of the reactor are required to be maintained in order to minimize the needless consumption of hydrogen.In the present interval of time, the Pr Ox is the main technique in the development of hydrogen. Generally, the catalysts are noble metals, for example, platinum, ruthenium, or rhodium supported on Al_2O_3.

At the same point of time, hydrogen (H_2) is purified with the help of the alternative approaches, which is named as pressure-swing adsorption, cryogenic distillation, and membrane technologies, which can make sure the essential purity of hydrogen (that is near about 98% to 99%). The most advantageous gas-purification method is the pressure-swing adsorption for its high efficiency (which is more than 99.99%) and flexibility as well.

3.7. HYDROGEN STORAGE

Indeed, the most obvious difficult tasks that are linked with the use of hydrogen gas as a fuel are its low density, high volume, and flammability. In order to use hydrogen (H_2) as a replacement for fossil fuels in cars that are powered with the help of fuel cells, the hydrogen fuel must be safely delivered in the amount which is suitable and fulfill the driving-range necessities of the typical driver.

In addition to this, as long as this amount of hydrogen fuel to the vehicle must be provided without taking up a lot of space or adding excessive weight. Given hydrogen's properties, this is not trivial.

The content of the energy of a kilogram of hydrogen (H_2) is near about equivalent to 4 L of gasoline (that is based on its lower heating value (LHV)), and for the same amount of energy, hydrogen gas takes up about three thousand times more volume (Armaroli and Balzani, 2011; U.S. Department of Energy (DOE) and Office of Energy Efficiency and Renewable Energy, 2009).In simple words, it would take more than 3400 L (near about 900 U.S. gallons) in order to substitute 4.5L (which is approximately 1 gal) of gasoline at standard temperature and pressure (Graetz, 2009). In this way, not only it is required to find out a different approach or different tactic in order to produce huge amount of hydrogen fuel, but also, there is a requirement to find out an approach in order to store the hydrogen safely and compactly, in such a way that the hydrogen can be delivered as per the requirements.

Hydrogen storage is extensively well thought out as the biggest obstacle or challenging task to the implementation of the hydrogen economy and is an area of explosive research that goes hand in hand with the development of the hydrogen fuel cell (Figure 3.12).

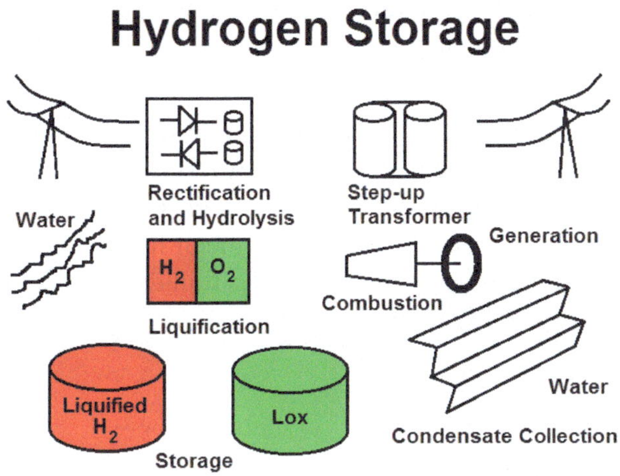

Figure 3.12. An illustration of hydrogen storage.

Source: Image by Wikimedia commons.

Hydrogen can be stored by physical means as well as by chemical means. In order to store the hydrogen gas with the help of physical compression to the liquid state is not commercially feasible for the transportation sector. According to that, the boiling point of liquid hydrogen is–253°C. In this way, the compression is very costly and boil-off is a big problem.

In addition to this, even liquid hydrogen has a lower energy density by volume when it is compared with a liquid hydrocarbon, for example, gasoline.

The compression of hydrogen gas up to 71 MPa (or near about 700 atm) is feasible but, at the same point of time, it is very expensive in nature and presents several numbers of various engineering problems.

The storage tank is required to be larger as compared to a gasoline tank. The shape of the storage has to be cylindrical in order to bear the high pressure (making it challenging task to design into the vehicle), and several numbers of various materials fail to prevent hydrogen diffusion, posing a potentially dangerous leakage problem (Armaroli and Balzani, 2011).

Chemical solutions to the hydrogen storage problem (materials-based hydrogen storage or chemical hydrogen storage (CHS) would ideally provide hydrogen fuel in a low-pressure, controlled manner that permits refueling.

Irrespective of the chemical process is involved, the storage unit, once depleted, should be able to be refilled in such a way, that the vehicle can be powered over and over again with the hydrogen-generated electricity of the fuel cell.

There are three general chemical tactics or ways to the challenges of hydrogen storage are presented in the above sections. The first is metal organic frameworks (MOFs) in order to absorb (as well as release when it is required) hydrogen gas. Second is the metal hydride as the carriers of the hydrogen, and the third is the use of the liquids that are enriched with hydrogen, as surrogate hydrogen storage.

3.8. CONCLUSION

To conclude this chapter, there is tremendous amount of research being pursued in the direction of the development of the technologies that are being used in order to produce the hydrogen. There are several numbers of various methodologies in order to produce hydrogen. Some of them are very expensive, but at the same point of time, there are couples of techniques that are cost effective in the production of the hydrogen.

The production of the hydrogen can be done with several numbers of tactics, such as SR, POX, ATR, water gas shift, and Pr Ox. Out of which, the SR is most cost effective.

The production of the hydrogen can also be done with the help of the fossil fuels and same, with the help of the electric current. But again, production of the hydrogen is very costly and it is being produced with the help of the electricity as compared to the fossil fuels.

In order to reduce the dependence on fossil fuels, there are meaningful developments in other technologies of production of hydrogen (H_2) from renewable resources such as biomass and water are considered.

It is very significant to note that hydrogen can be produced from a huge variety of feedstock that is available almost everywhere on the planet. There are several numbers of various processes under development with minimal environmental impact.

Development of these several technologies can reduce the world's dependence on fuels that come mainly from the unstable regions. The production of the "in-house" hydrogen can elevate both national energy as well as the economic security.

The potential of hydrogen to be produced from an extensive variety of feedstock and with the help of an extensive variety of processes can make each and every region of the world to be able to produce much of its own energy. It is quite clear that as the technologies develop and enhance, the hydrogen may prove to be the most ubiquitous fuel available on the planet Earth.

REFERENCES

1. ALL Power Labs, (2020). The Five Processes of Gasification-ALL Power Labs. [online] ALL Power Labs. Available at: http://www.allpowerlabs.com/gasification-explained (accessed on 4 July 2020).
2. Carpenter, N., (2014). Chemistry of Sustainable Energy. [online] Boca Raton, Florida: CRC Press. Available at: https://www.routledge.com/Chemistry-of-Sustainable-Energy/Carpenter/p/book/9781466575325 (accessed on 4 July 2020).
3. Hiden, A., (2020). What Is Catalysis? [Online] Hiden Analytical. Available at: https://www.hidenanalytical.com/what-is-catalysis/ (accessed on 4 July 2020).
4. Hugh, S. T., (2020). Catalysis | Chemistry, Classification, and Chemical Reactions. [Online] Encyclopedia Britannica. Available at: https://www.britannica.com/science/catalysis (accessed on 4 July 2020).
5. Hugh, S. T., (2020). Catalysis-Classification of Catalysts. [Online] Encyclopedia Britannica. Available at: https://www.britannica.com/science/catalysis/Classification-of-catalysts (accessed on 4 July 2020).
6. Hugh, S. T., (2020). History of Catalysis. [Online] Encyclopedia Britannica. Available at: https://www.britannica.com/science/catalysis#ref277254 (accessed on 4 July 2020).
7. Hugh, S. T., (2020). Homogeneous Catalysis | Chemistry. [Online] Encyclopedia Britannica. Available at: https://www.britannica.com/science/homogeneous-catalysis (accessed on 4 July 2020).
8. Kalamaras, C., & Efstathiou, A., (2013). Hydrogen Production Technologies: Current State and Future Developments. Conference Papers in Energy, [Online] 2013, pp. 1–9. Available at: https://www.hindawi.com/journals/cpis/2013/690627/ (accessed on 4 July 2020).
9. Manoelle, L., Marie-Noëlle, S., Alain, B., Bernard, T., & Cécile, M., (2015). Hydrogen Production. [Online] PlanèteÉnergies. Available at: https://www.planete-energies.com/en/medias/close/hydrogen-production (accessed on 4 July 2020).
10. Maria G., Pin-Ching, M., Richard, F., Todd, D., & John, T., (n.d.) Hydrogen Production and Delivery | Hydrogen and Fuel Cells | Hydrogen and Fuel Cells | NREL. [Online] Nrel.gov. Available at: https://www.nrel.gov/hydrogen/hydrogen-production-delivery.html (accessed on 4 July 2020).
11. Mustain, W., Kohl, P., Palomo, J., Wang, R., Yong, X., Gao, W., Lin, X.,

et al., (2020). Catalysis-Latest Research and News | Nature. [Online] Nature.com. Available at: https://www.nature.com/subjects/catalysis (accessed on 4 July 2020).

12. National Energy Technology Laboratory, (n.d.) 5.1. Gasification Introduction. [Online] netl.doe.gov. Available at: https://www.netl.doe.gov/research/Coal/energy-systems/gasification/gasifipedia/intro-to-gasification (accessed on 4 July 2020).
13.

Chapter 4 Fuel Cells and Solar Photovoltaic

CONTENTS

The chapter of fuel cells and the solar photovoltaic explains the basic fundamentals and principle of fuels cells such as how do fuel cells work, and how it converts chemical energy into the usable electrical energy. This chapter also explains the development or the designing of fuel cell systems. This chapter also provides a highlight on the several numbers of various fuel cells such as alkaline fuel cells, phosphoric acid fuel cells, molten carbonate fuel cell, and solid oxide fuel cell.

This chapter explains the basic significance of the solar photovoltaic. This chapter explains the concept of the solar photovoltaic energy such as benefits of the photovoltaic energy and the difference between the peak and nominal capacity in photovoltaic installations. This chapter also addresses the basics of the solar photovoltaic technology such as silicon solar cells, thin file solar cells, III-V solar cells, and next generation solar cells.

This chapter also explains the working of the solar photovoltaic system, components of the solar photovoltaic system such as solar panels, inverters, racking, along with the other components. This chapter also includes the benefits of the using solar photovoltaic panel such as cost savings, reduced carbon emission, tax credits, and clean energy. This chapter also provides highlights on the advantages of the solar photovoltaic as well as the disadvantages of the solar photovoltaic.

4.1. INTRODUCTION

Not similar to the conventional technologies of combustion that burn fuel, fuel cells went through a chemical process in order to transform the fuel which is rich with content of hydrogen into the electricity. Fuel cells do not require to be eventually recharged as batteries, but in spite of continue to develop electricity as long as a fuel source is provided.

A fuel cell is made up with the help of an anode, a cathode, and a membrane of the electrolyte. A fuel cell operates with the help of passing the hydrogen via an anode of a fuel cell and the oxygen via cathode. At the site of anode, the molecules of the hydrogen are separated into the electrons and protons. The protons pass through the membrane of the electrolyte.

On the other hand, the electrons are forced through a circuit, developing an electric current and excess heat. At the site of cathode, the electrons, protons as well as oxygen enters the fuel cell. With respect to the anode site, the molecules of the hydrogen are separated into the protons and electrons.

The protons pass across the membrane of the electrolyte, on the contrary, the electrons are forced through a circuit, as a result, it develops an electric current and extra amount of heat. At the site of cathode, the electrons, protons, along with the molecules of the oxygen altogether in order to produce molecules of the water.

Because of their higher efficacy, the fuel cells are very clean, with their only by-products being electricity, extra amount of water as well as the extra amount of heat. Moreover, as the fuel cells do not have any kind of moving parts, they only function very silently.

Fuel cells are also scalable in nature. In simple words, the individual fuel cells can be accumulated on one another in order to produce stacks, as an outcome, these stacks can be integrated into the larger systems.

The system of the fuel cell can fluctuate extensively with respect to their size and their power, from portable systems for the smartphone battery charging to the substitute of combustion engine for electric vehicles, to large-scale, multi-megawatt installation offering the flow of electricity directly to the utility grid.

Fuel cell could be any sort of devices which transform the chemical energy of a fuel directly in to the form of electricity with the help of an electrochemical reactions. A fuel cell is similar to a battery in several numbers of various aspects, but it has only ability to supply the energy which is in the form of electricity over a much longer interval of time.

And the reason behind that is that a fuel cell is non-stop supplied with fuel and air (or with the oxygen) with the help of an external source. On the other hand, a battery which consisting only of a limited amount of fuel material along with the oxidant that are exhausted on using (Figure 4.1).

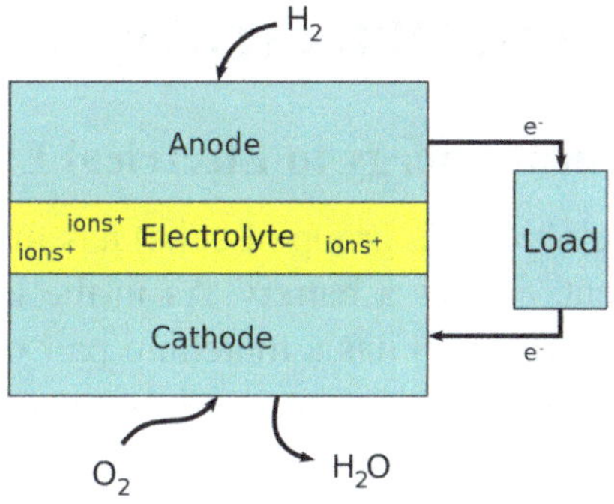

Figure 4.1. An illustration of fuel cell.

Source: Image by Wikipedia.

For this reason, the fuel cells have been used for several numbers of years in space probes, in manned spaceships, along with the satellites. All across the globe, hundreds, and thousands of the stationary fuel cell systems have been installed in the utility power plants, schools, office buildings, hotels, along with the hospitals for both primary sources of power as well as backup power.

Several numbers of waste treatment plants also use the technology of fuel cells in order to develop power with the help of methane (CH_4) gas which is produced by decomposing of garbage.

Several numbers of different municipalities in the United States, Europe, and Japan lease vehicles that operate over the fuel cells for public transportation and for the use by service personnel. Personal owned vehicle that works with the help of fuel cell was first sold in Germany in the year 2004.

The government of the United States of America, along with the various other governments, most notably California, have launched the programs in order to boost the development as well as the application of the hydrogen fuel cells in transportation as well as the other applications.

Never the less, the technology of fuel cells has proven to be workable, attempts to make it competitive in commercial areas has been comparatively less successful. And the reason behind this concern with the explosive power of hydrogen, the comparatively low energy density of hydrogen, as well as the high cost of the platinum catalysts used in order to develop the electric current with the help of splitting the electrons from the atoms of hydrogen.

4.2. PRINCIPLES OF OPERATION

4.2.1. From Chemical Energy to Electrical Energy

A fuel cell (typically, which is a group of cells) has fundamentally the same type of the components just as a battery. As in the latter, each, and every cell of the system of a fuel cell has a matching pair of electrodes. There is the anode, which supplies the electrons, and on the other hand, there is the cathode, which absorbs the electrons.

Both of the electrodes must be immersed in and separated with the help of an electrolyte, which can be in the form of liquid or solid, but which must in either case conduct ions between the electrodes in order to fulfill the chemistry of the system.

A fuel, for example, hydrogen, is supplied to the anode. At the site of anode, the hydrogen is being oxidized; as a result, it produces the ions of hydrogen along with the electrons. An oxidizer, for example, oxygen, is supplied to the cathode. At the site of cathode, the ions of the hydrogen from the anode absorb electrons from the latter and react with the oxygen in order to generate water.

The main difference among the respective levels of energy at the electrodes (which is an electromotive force) is the voltage per unit cell. The amount of the electric current that is available to the external circuit relies on the chemical activity as well as the amount of the substance that is supplied in the form of fuels.

The process of producing the current endures for as long as there is a supply of reactants, with respect to the electrodes and the electrolytes of a fuel cell, not similar to those, that are used in regular batteries. These are designed in such a way that they remain unchanged by the chemical reaction.

A practical fuel is more and more a complicated system. The practical fuel system provide assistance in order to increase the activity of the fuel, pumps, and blowers, fuel-storage containers, and a variety of sophisticated sensors and controls with which to monitor and regulate the operation of the system. The operating potential as well as the life span of each and every of the design of this system can limit the performance of the fuel cell.

With respect to the other electrochemical systems, the function of the fuel cell is relying on the temperature. The chemical activity of the fuels and the value of the activity promoters, or the catalysts, are decreased with the help of the temperature for example, zero degree Celsius or 32°F.

On the contrary, very high temperatures enhance the activity aspects but they can decrease the functioning life time of the electrodes, construction materials, blowers, as well as the sensors. In this way, each, and every type of the fuel cell has an operating-temperature design range, and a significant departure from this range is likely to decrease both, capacity of the fuel cell as well as the lifetime of the fuel cells.

A fuel cell, for example, a battery, is fundamentally a device which has high effectiveness, not like the internal combustion machines. Talking about the internal combustion machines, the fuel is burned, and the gas is expanded in order to perform the work. On the contrary, the fuel cells transform the chemical energy directly in the form of electrical energy.

Due to this fundamental property, fuel cells can transform the fuels in to the energy, which is usable in nature, at an efficacy as high as 60%. On the other hand, the internal combustion machine or internal combustion engine, the effectiveness is limited to near about 40% or less. In simple words, the high efficacy means that very less fuel and a smaller storage container are required for a fixed energy requirement.

Due to this reason, fuel cells are an attractive power supply with respect to the space mission of limited interval of time and for other circumstances as well, where the fuel is very costly in nature and a challenging task to supply.

In addition to this, fuel cells do not emit harmful or toxic gases like nitrogen dioxide and create virtually no noise through the course of operation, which gives a practical and good reason to make them contenders for local municipal power generation stations.

A fuel cell can be designed in order to function reversibly as well. In simple words, a hydrogen-oxygen cell that produces water as a product can be made in order to develop regenerate hydrogen as well as oxygen. These sorts of regenerative fuel cell are consists not only a revision of electrode design, but apart from that, it also introduces the special means for separating the product gases.

Sooner or later, the power modules that are consisting of this kind of high efficacy fuel cells can be used in conjunction with large arrays of thermal collectors for solar heating or other solar energy systems can be utilized in order to keep the cost of the energy cycle lower in longer-lived equipment.

All across the globe, major companies that deal with the automobiles and firms that manufactures with the help of electrically driven machines have announced their intention to generate or use fuel cells commercially in the coming interval of time.

4.2.2. How Do Fuel Cells Work?

The main cause of a fuel cell is to generate the electric current that can be provided outside the cell in order to perform the work, for example, as providing power to an electric motor or illuminating a light bulb or a city. Because of the way electricity behaves, this current returns to the fuel cell, completing an electrical circuit.

There are several numbers of various types of fuel cells and each of them functions differently when they are compared to each other. With respect to general terms, hydrogen atoms enter a fuel cell at the anode where a chemical reaction strips them of their electrons. At that interval of time, the atoms of the hydrogen are "ionized," and carry a positive electric charge.

The electrons, which are, by default negatively charged, supply the current with the help of the wires in order to perform the work. If Alternating Current (AC) is required, the Direct Current (DC) output of the fuel cell must be routed with the help of a conversion device, which is known as an inverter.

As the oxygen enters in the fuel cell at the cathode end, in some cell types, the oxygen integrates with the electron that are returning from the electrical circuit and hydrogen ions that have travelled with the help of the electrolyte from the anode. In the other cell types the oxygen picks up the electrons and then travels with the help of the electrolyte towards the direction of anode. At the site of anode, it integrates with the ions of the hydrogen.

The electrolyte plays the main role. Electrolyte must allow only the suitable ions to pass between the anode and the cathode. If the free electrons or the other substance could travel with the help of the electrolyte, they would interrupt or slowdowns the rate of the chemical reaction.

Whether they integrate at the site of the anode or the site of the cathode, together hydrogen and the oxygen produce water, which drains from the cell. As long as a fuel cell is supplied with the hydrogen and oxygen, it will generate the electricity.

Even better, from the time the fuel cells generate the electricity with the help of the chemical reaction, in spite of, with the help of the combustion, they are not subject to the thermodynamic laws that limit a traditional power plant.

In this way, the fuel cells are more effective, or they are having efficacy in extracting the energy from a fuel. Waste heat, which produced from some cells, can also be harnessed, provide assistance in order to increase the efficacy of the system further.

4.3. DESIGNING FUEL CELL SYSTEMS

Due to the reason that, a fuel cell not-stop produces electric current from the fuel, it has several numbers of various output properties that are identical to those of any other DC generator system. A DC generator system can be functioned in either of two ways from a planning point of view:

1. Fuel can be burned inside the heat engine in order to drive the electric generator, which makes power available and the flow of the current, or
2. Fuel can be transformed, in order to form which is appropriate with respect to a fuel cell, after which, it generates electricity or power directly.

An extensive range of liquid fuels as well as the solid fuels can be used for the system of a heat engine. On the other hand, hydrogen, methanol, and reformed natural gas (for example, methane that has been transformed into the gas which is rich in hydrogen content) are the primary fuels that are available for current fuel cells.

If fuels, for example, natural gas must be transformed or converted into the composition for a fuel cell, the net efficacy of the fuel cell system is decreased, and much of its efficiency advantage is lost. These sorts of "indirect" fuel cell system would still show an effective advantage as high as 20%.Never the less, in order to be competitive with the modern thermal generating plants, a fuel cell system must acquire a good design balance with low internal electrical losses, electrodes that are corrosion resistance, an electrolyte made up of constant composition, low catalyst costs, and ecologically acceptable fuels.

The first technical difficulty that must be overcome in order to produce practical fuel cells is to design and bring together an electrode that permits the fuel which can be present in the form of gas or liquid, to contact a catalyst and an electrolyte at a group of solid sites that do not alter very rapidly.

In this way, a three-phase reaction situation is generally taking place on the electrode, and that must also serve as a conductor of the electric current. These can be offered with the help of thin sheets that have:

1. A waterproof layer, which is generally made up with the help of polytetrafluoroethylene (Teflon),
2. An active layer of a catalyst (for example, platinum, gold, or more complicated organometallic compound on a carbon base), and
3. A conducting layer, in order to carry the current which is being generated in or out of the electrode.

If the electrode floods with the electrolyte, the rate of the operation will become very slow at best. If the fuel breaks through to the electrolyte side of the electrode, the compartment of the electrolyte can become filled with gas or vapor, inviting an explosion should the oxidizing gas also reach to the compartment or the fuel gas enter the oxidizing gas compartment.

In simple words, in order to maintain the stable operation in a working fuel cell, careful design, construction, and pressure control plays a very important role. And the reason behind this is that, the fuel cells have been used on Apollo lunar flights along with all other United States orbital manned space missions (for example, those of Gemini and space shuttle), it is a proof that all of the three necessities can be fulfilled reliably.

Offering a fuel cell support system of pumps, blowers, sensors, along with the controls in order to regulate or maintain the fuel rates, the load of the electric current, gas pressure as well as liquid pressure, and temperature of the fuel cell remains a major difficulty in engineering design.

Considerable amounts of enhancements in the service life of these various components under adverse setting would contribute to an extensive application of the fuel cells.

4.4. TYPES OF FUEL CELLS

There are several numbers of various fuel cells have been developed. Typically, the fuel cells are categorized on the basis of the electrolyte that is being used the electrolyte identifies the operating temperature of a system and in part the kind of fuel that can be used.

4.4.1. Alkaline Fuel Cells

The definition of the alkaline fuel cells is that, the devices which have a liquid solution of hydroxide or potassium hydroxide as the electrolyte. The fuel is always hydrogen gas, with oxygen (or oxygen in air) in the form of oxidizer.

Never the less, zinc or aluminum can also be used in the form an anode, if the by-product oxides were effectively removed and the metal fed continuously in the form of a strip or in the form a powder. Generally, the fuel cells operate or functions at the temperature which is less than 100°C or two hundred-and twelve-degrees Fahrenheit and are made up with the help of metal and specific sort of plastics (Figure 4.2).

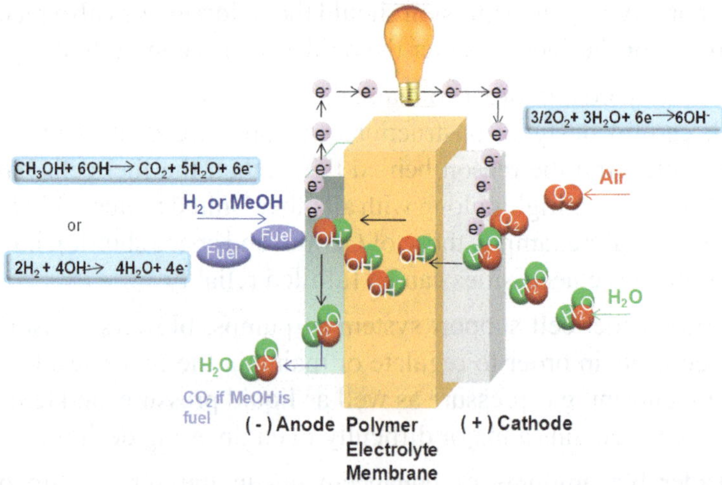

Figure 4.2. An illustration of alkaline fuel cell.

Source: Image by Wikimedia commons.

Electrodes are made up with the help of carbon along with a metal, like nickel. Water is being used as reaction product, and it must be removed from the system. Generally, the removal of water takes place is with the help of evaporation from the electrolyte either through the electrodes or in a separate evaporator.

The operating support system shows a significant challenge with respect to the design of fuel cells. The strong, hot alkaline electrolyte attacks most plastics and tends to pierce or infiltrate the structural seams and joints. This problem has been overcome, never the less, and alkaline fuel cells are used on the United States space shuttle orbiters.

The overall effectiveness ranges from near about 30% to 80%, and that relies on the fuel and oxidizer and on the basis of the calculations.

4.4.2. Phosphoric Acid Fuel Cells

These kinds of cells have an orthophosphoric acid electrolyte that permits operation up to near about 200ºC or 400ºF. Phosphoric Acid Fuel Cells can use a hydrogen fuel, which is soiled with the help of the carbon dioxide (CO_2) and an oxidizer of air or oxygen.

The electrodes of phosphoric acid fuel cells are consisting of catalyzed carbon and are arranged in pairs that are set back-to-back in order to create a series generation circuit.

The farming structure for this assembly of cells is made up with the help of graphite, which remarkably increases the cost of that particular fuel cell. The higher temperature and aggressive hot phosphate produce issues with the structural design, more specifically in the joints, supporting pumps, and sensors.

Phosphoric acid fuel cells have been proposed and tested on a limited scale for local municipal power station and for remote-site generators.

4.5. MOLTEN CARBONATE FUEL CELL

The MCFC, which is the short form for molten carbonate fuel cell (MCFC), uses a liquid or molten mixture of alkali metal carbonates in the form of the electrolyte and the ions of the carbonate acts as the charge carrier. The sites of the anode are made up with the help of a porous, sintered nickel chromium alloys (Ni-Cr Alloys) or nickel aluminum alloys (Ni-Al Alloys). On the other hand, the cathode sites are made up with the help of the nickel oxide.

The electrolyte is kept in a ceramic matrix of $LiAlO_2$ (Lithium aluminate). Generally, it is the mixture of lithium and potassium or lithium and sodium carbonates that produces a molten salt which is highly conductive in nature, at a very high temperature (near about 600°C to 700°C).

As these sorts of devices functions at high temperature, non-precious metals can be used in the form of catalysts at the anode and the cathode, in this way, it helps in reducing the cost of it. The ions of the carbonate in the presence of the hydrogen are needed in order to produce water and CO_2. On the other hand, at the site of cathode, the CO_2 is then reduced and reacts with the oxygen in order to reform the carbonate.

Not similar to alkaline, phosphoric acid, and PEMFCs (Proton-exchange membrane fuel cells), MCFC does not necessitate an external reformer in order to transform the fuel like natural gas into the hydrogen (Figure 4.3).

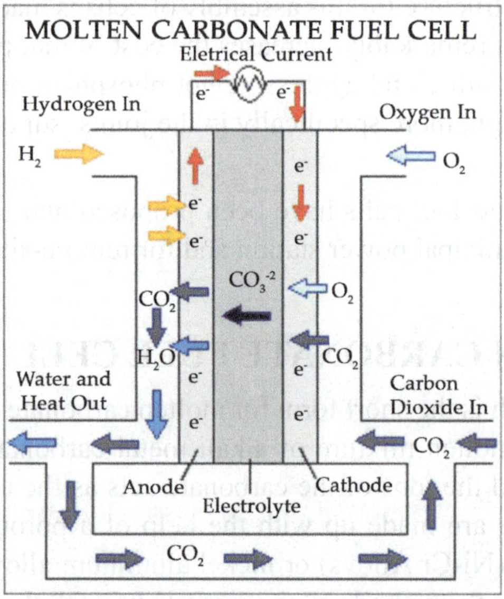

Figure 4.3. An illustration of molten carbonate fuel cell.

Source: Image by Wikimedia commons.

At the high temperature, at which MCFC operates, CH_4 along with the other hydrocarbons that are light in weight are transformed into the hydrogen within the fuel cell itself with the help of the process referred to as internal reforming, which again, helps in reducing the cost of MCFC.

Never the less, as MCFC operates on the temperature which is high enough in order to maintain the electrolyte molten, they are more suited for stationary applications as opposed to the mobile application. Moreover, the high temperature as well as the corrosive electrolyte increase the rate of the breakdown of the component and corrosion as well, which as an outcome, significantly reduces the life of the fuel cell.

4.6. SOLID OXIDE FUEL CELL

SOFC, which is the short form of Solid Oxide Fuel Cells (SOFCs) are a kind of fuel cell that uses a solid oxide material in the form of the electrolyte. SOFCs uses a solid oxide electrolyte in order to conduct the negative oxygen ions from the cathode to the anode. The electrochemical oxidation of the ions of the oxygen with the hydrogen or carbon monoxide (CO), in this way, takes place on the site of anode.

SOFCs function at very high temperatures (that is near about 500°C to 1000°C). At these high temperatures, SOFCs do not necessitate costly platinum catalyst, as it presently important with respect to the lower-temperature fuel cells, like Proton Exchange Membrane Fuel Cells (PEMFC) and are not vulnerable to the CO catalyst poisoning (Figure 4.4).

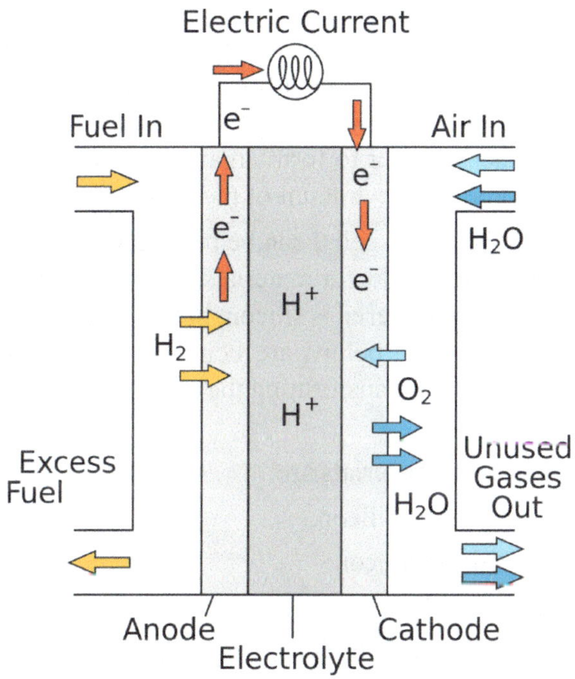

Figure 4.4. An illustration of solid oxide fuel cell.

Source: Image by Wikimedia commons.

Never the less, the vulnerability to sulfur poisoning has been extensively noted and the sulfur must be removed before entering the cell with the help of the adsorbent beds or with the help of other means as well.

SOFCs have an extensive variety of application from use an auxiliary power unit in vehicles to stationary power generation with outputs from hundred Watts to two Mega Watts. Higher operating temperature makes SOFCs an appropriate contender for the application with the heat engine energy recovery device or integrated heat and power, which increases the overall effectiveness of the fuel.

4.7. SOLAR PHOTOVOLTAIC

Solar photovoltaic (PV) is a technology which transforms the light coming from the sun or solar radiations into DC electricity with the help of the semiconductors. When the sunlight or solar radiation hits the semiconductor that is placed within the PV cells or PV cell, the electrons are freed and hence produce the electricity.

Generally, the technology of the solar PV is used on a panel (that are solar panels). PV cells are generally found linked to each other and mounted on a frame, which is known as a module. Several numbers of various modules can be wired together in order to form an array, which can be scaled up or drown in order to produce the amount of the power that is required.

Photovoltaic cells (or PV cells) can be made up with the help of several numbers of various semiconductor materials. In the present interval of time, the most commonly used material is silicon, but other materials, for example, the materials that are listed below, are being tested and used in order to elevate the effectiveness of transforming the sunlight or solar radiations into the electric current.

The most commonly materials are:

1. Monocrystalline Silicon;
2. Polycrystalline Silicon;
3. Amorphous Silicon;
4. Cadmium Telluride (Cd-Te); and
5. Copper Indium Gallium Selenide (CIGS).

Figure 4.5. An illustration of solar photovoltaic.

Source: Image by needpix.com.

In the present interval of time, all across the globe, near about 90% of the photovoltaic (PV) technologies are based on some kind of silicon. In the year of 2011, near about 95% of all shipments by the United States manufacturers to the residential sector were crystalline silicon solar panels (Figure 4.5).

The biggest difference among the technologies is the material that is being used in order to generate the electric current with the help of the sunlight. Each and every type of the material has various kinds of characteristic or properties, as an outcome, different applications and the efficacy. Generally speaking, the effectiveness of the solar PV technologies fluctuates, and that ranges from 6% to 18% at the present interval of time.

4.8. SOLAR PHOTOVOLTAIC ENERGY

Solar PV energy or PV solar energy directly transforms the sunlight or solar radiation into the electric current, with the help of the technology that is based on the PV effect. When the radiation from the sun hits one of the faces of the photoelectric cell (several numbers of PV make up a solar panel), it generates an electric voltage differential between both faces that makes the electrons flow between one to the other, producing the electricity (Figure 4.6).

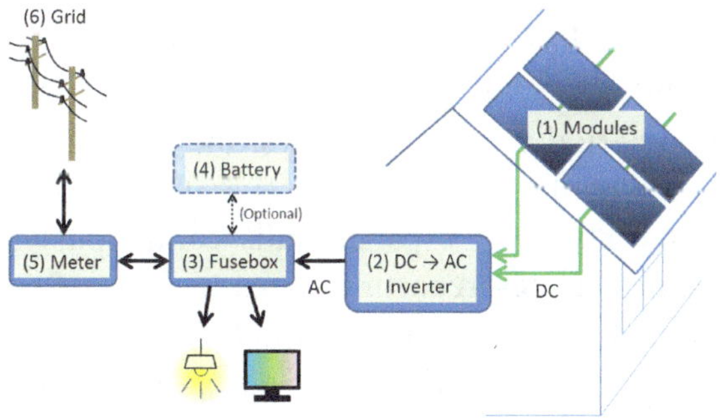

Figure 4.6. An illustration of solar photovoltaic energy.

Source: Image by Wikimedia commons.

There are three kinds of solar panels that are being used all across the globe. The first and foremost is the PV, which is the generator of electricity that is being supplied to the homes. Second type of solar panel is thermal, which is installed on houses in order to receive the light coming from the sun directly. And the third and last is thermodynamic, which functions in several kinds of weather settings, that is at night, when it is raining or cloudy.

When the PV technology first began, the PV was used in order to provide electricity to the satellites. According to the Spanish Association of Renewable Energy Producers (APPA), development of PV panel sped up in the time interval of 1950s and now has become a substitute to the application of the fossil fuels.

4.8.1. Benefits of Photovoltaic Energy

The electric current or electricity that is produced with the help of the PV panels is inexhaustible and does not pollute, and in this way, contributes to sustainable development as well as favoring local employment. In the similar manner, it can be exploited in two various ways. The first is sold to the electricity grid or utilized in isolated locations where there is no traditional electricity network.

In this way, it is a particularly efficient system with respect to the remote and rural regions, which cannot be reached with the help of the electric power lines, or they are difficult or very expensive in order to install, and in nations receiving several numbers of hours of sunlight per year.

The cost of installing and maintaining the solar panels, whose average useful life is more than three hundred years, has come down considerably in the time period of recent years, as PV technology has developed. It necessitates an initial investment and small operational budget, but, once the PV system is installed, the "fuel" is free and is available for life.

Over the time period of next ten years, the solar energy will become the cheapest source of the electricity in several numbers of various parts of the world as the cost of PV panels falls continuously, guarantees the International Business Times, reporting on research carried out with the help of the German think tank Agora Energiewende.

From the time period of 1980s, panels for producing the electric current with the help of the sunlight or solar radiation have been falling in price at near about 10% with each passing year. With respect to this rate, the technology will be able to provide near about 20% of energy requirement by the year of 2027, all across the globe.

The reputed business magazine Fortune mentioned a study which concludes that the solar energy will be providing more and more jobs when it is compared with the oil sector by the end of the year 2019. In the United States, as per the Solar Foundation, new employment in 2019 in the solar energy sector for the third year running.

The global market for the PV energy is anticipated to increase by near about 20% in the coming time interval of two years, adding at least 60 GW in the year of 2016, and more than 70 GW in the year of 2017, as per the study done by the PV Market Alliance (PVMA). The same organization having a faith that demand will last to grow, and the main reason behind that is the strength growth in India, China, the United States, and various other emerging markets.

Certainly, as per the reports of the PVMA, from the data that is being collected on its renewable energies website, the combination of 51 GW in the year of 2015, all across the globe, meant the sector set a new record for installed PV capacity, with Asia the top continent in this context.

4.8.2. Difference between Peak and Nominal Capacity in Photovoltaic Installations

With respect to the PV solar panels, the peak capacity means the amount of Kilo Watt installed. On the other hand, the nominal capacity means the capacity of the inverter (the electrical equipment that transforms the electricity which is developed with the help of the panels so it can be utilized by the consumers).

According to the definition, the nominal capacity is the limit of the plant (that an individual cannot generate more than the inverter can transform). Never the less, PV facilities always install a peak capacity greater than the nominal. There is requirement of more panels in order to make sure that the 100% capacity of the inverter is utilized.

A solar PV plant will be well designed if the peak capacity, the installed panels, that makes sure the inverter can function at 100% capacity whenever and wherever it is needed.

4.9. SOLAR PHOTOVOLTAIC TECHNOLOGY BASICS

Photovoltaic (sometimes shortened a PV) gets its name from the process of converting light (that is packets of protons) into electricity (or voltage), which is known as the PV effect. This phenomenon was first exploited in the year of 1954 with the help of the scientists at Bell Laboratories who developed a working solar cell which is made up with help of silicon and developed an electric current when it is exposed to the solar radiations or sunlight.

Solar cells were soon being used in order to provide power to the space satellites and smaller items, for example, watches, and calculators. In the present interval of time, the electricity from the solar cells has become cost-competitive in several numbers of various areas and PV systems are being used at large scales in order to provide power to the electric grids.

4.9.1. Silicon Solar Cells

In the present interval of time, the enormous majorities of the solar cells are made up with the help of the silicon and offer both reasonable cost as well as good efficacy (which are the rate at which the solar cell transforms the solar radiations into the electric current).

These kinds of cells are generally assembled into a larger module that can be installed on the roofs of residential as well as commercial buildings or used on the ground mounted racks in order to create huge, utility-scale systems.

4.9.2. Thin-Film Solar Cells

Apart from silicon solar cells, the other commonly used PV technology is known as the thin-film solar cells. And the reason behind this is that they are made up with the help of very thin layers of semiconductor material, for example, cadmium telluride or copper indium gallium diselenide. The thickness of these kinds of cell layers is only a few micrometers, which is several millionths of a meter.

Thin-film solar cells can be flexible and lightweight in nature, making them ideal for portable applications, for example, in a backpack of a soldier, or for use in the other products such as windows that produce electricity from the solar radiations.

There are some sorts of thin-film solar cells that also benefit from manufacturing techniques that necessitates less energy and are very easy to scale-up when it is compared to the manufacturing techniques that are needed by silicon solar cells.

4.9.3. III-V Solar Cells

The third type of PV technology is named after the elements that compose them. III-V solar cells are primarily made up with the help of the elements in Group-III, for example, gallium, and indium. Group-V, for example, arsenic, and antimony of the periodic table.

Typically, these kinds of solar cells are much more costly with respect to the manufacture when it is compared to the other technologies. But they transform the solar radiation into the electricity at much higher effectiveness.

And due to this reason, these solar cells are every so often used in the satellites, unmanned aerial vehicles, and the other application that necessitates a high ratio of power-to-weight.

4.9.4. Next-Generation Solar Cells

Solar cell researchers at National Renewable Energy Laboratory (NREL) and elsewhere are also pursuing several numbers of various new PV technologies, for example, solar cells made up with the help of organic materials, quantum dots, and hybrid organic-inorganic materials (also known as perovskites).

These next-generation technologies can provide lower costs, greater ease of manufacture, or the other benefits. In addition to this, research will see if these promises can be realized.

4.10. HOW DO SOLAR PHOTOVOLTAIC SYSTEMS WORK?

The light coming from the sun, or solar radiation, which is made up with the help of packets of energy, which is known as photons that falls onto a solar panel and develops the electric current with the help of process which is known as the PV effect.

Each and every panel produces a comparatively small amount of energy but can be associated together with other panels in order to develop higher amounts of energy as a solar array.

The electricity which is generated from a solar panel (or also known as array) is in the form of the DC. Never the less, several numbers of various electronic devices that use DC electricity, which is consisting of the smartphone or laptop, they are designed in order to operate with the help of the electrical utility grid which provides (and it requires) alternating current (AC).

In this way, with respect to the solar electricity to be useful it must first be transformed from DC to the AC with the help of an inverter. This AC electricity from the inverter can then be used in order to power electronics locally or be sent on to the electrical grid for use elsewhere.

4.11. COMPONENTS OF SOLAR PHOTOVOLTAIC SYSTEM

Moreover, to the solar panels, there are other important components of a PV system which are generally referred to as the "balance of the system" or BOS. These several components (that are generally account for near about half of the cost of the system and most of the maintenance) can be consisting of inverters, racking, wiring, combiners, disconnects, electric meters, and circuit breakers (Figure 4.7).

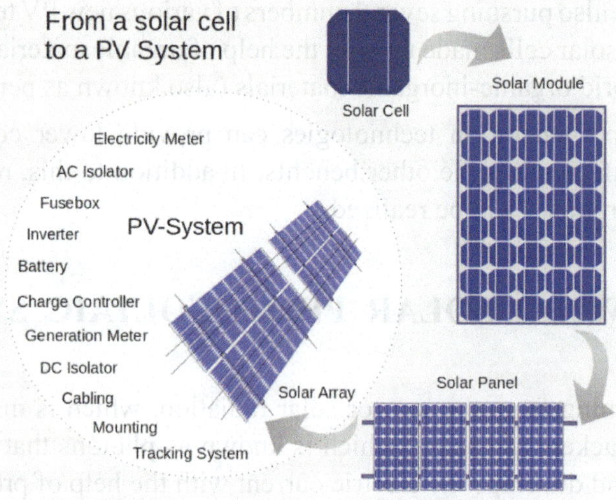

Figure 4.7. Components of the solar photovoltaic system.

Source: Image by Wikipedia.

4.11.1. Solar Panel

A solar panel is consisting of several numbers of various solar cells with semiconductor characteristics sum up within a material in order to protect it from the environment. These characteristics permit the cell in order to capture the light, or more particularly, the photons from the sun and transform their energy into the useful electricity with the help of a process which is known as PV effect.

On the other side of the semiconductor is a layer of conducting material which "gathers" the electric current that has been generated. The illuminated side of the panel also contains an anti-reflection coating in order to minimize the losses because of the reflection. The majority of the solar panels that produced all across the globe are made up with the help of crystalline silicon, which has a theoretical efficacy limit of near about 33% in order to transform the solar radiation into the electric current. Several numbers of other semiconductor materials along with the solar cell technologies have been developed that operate at higher efficiencies, but these come with a higher cost in order to manufacture.

4.11.2. Inverters

An inverter is an electrical device which allows the electrical current as DC and then transforms it into the AC. With respect to the solar energy systems, this means that the direct current or DC from the solar array is fed with the help of the inverter that transforms it into the AC.

This conversion is very important in order to function most of the electric devices or interface with the help of electrical grid. Inverters are very significant aspect with respect to all systems that works on the solar energy and are generally the most expensive element after the solar panels themselves.

Most of the inverters have conversion effectiveness of near about 90% or higher than that, and they are consisting of important safety properties which that include ground fault circuit interruption and anti-islanding. These shut down the PV system when there is a loss of grid power.

4.11.3. Racking

Racking refers to the mounting apparatus which fixes the solar array to the ground or rooftop. Generally, it is made up with the help of steel or aluminum; these apparatuses mechanically fix the solar panels in place with a high level of precision.

Racking systems should be designed in such a way that it withstands in extreme weather settings, for example, hurricane or tornado level wind speeds and/or high accumulations of snow. Apart from this, the other important characteristic of racking system is to electrically bond and ground the solar array in order to prevent the electrocution.

Generally, the rooftop racking systems come in two different variations, out of which, the first one is consisting of flat roof systems and the second one is pitched roof systems. With respect to the flat rooftops, it is common for the racking system in order to include weighted ballast to hold the array to the roof with the help of the gravity. Talking about the pitched rooftops, the racking system must be mechanically attached to the roof structure. PV systems which are mounted to ground can also use either ballast or mechanical anchors in order to fix the array with the ground.

Apart from this, some of the ground mounted racking systems also consisting of tracking systems which use motors along with the sensors in order to track the sun through the sky, increment in the amount of energy generated at a higher equipment as well as high cost to maintain them.

4.11.4. Other Components

The rest of the elements of a typical solar PV system is consisting of combiners, disconnects, breakers, meters, and wiring. A solar combiner, as the name suggests, it integrates two electrical cables or more than two electrical cables into one larger one.

Generally, the combiners are consisting of fuses in order to protect and it is used on all medium to large and utility-scale solar arrays. Disconnects are electrical gates or witches permits for manual disconnection of an electrical wire.

Disconnects, generally used on both sides of an inverter, which is known as "DC Disconnect" and "Alternating Current Disconnect" these devices offer electrical gaud or electrical isolation when an inverter requires to be installed or substituted. Circuit breakers or breakers safeguard the electrical systems from over current or surges.

These are designed in such a way that, they trigger automatically when the current reaches a predetermined amount, breakers can also be operated manually, which acts as an additional disconnect. An electric meter gauges the amount of the energy that passes through it and is generally used with the help of electric utility companies in order to gauge and charge customers.

With respect to the solar PV systems, a special bi-directional electric meter is used in order to gauge both the incoming energy from the utility, as well as the outgoing energy from the solar PV system. In the end, the wiring or the electrical cables carries the electric energy or electric current from and between each component and must be properly sized in order to carry the current.

When the wiring exposed to the solar radiation, it must have protection against the Ultraviolet exposure, and wires carrying DC every so often necessitates the metal shielding for added protection.

4.12. BENEFITS OF USING SOLAR PHOTOVOLTAIC

There are several numbers of benefits of using the solar energy. Most of the benefits are move around the cost-effectiveness of the solar energy as well as environmental benefits. A list of most prominent benefits of using the solar energy that is produced with the help of the PV cells is mentioned below:

4.12.1. Cost Savings

The time, the sun is free, the solar system will cost some money. But, the cost savings come in once an individual or a person have paid for it all and he or she reaping the benefits of not having to live entirely off the grid and depending on the local utility company, and the inevitable increasing in the cost of the electricity.

4.12.2. Reduced Carbon Emissions

Apart from that, another benefit is related to the environment. When an individual or a person uses the solar energy, less carbon emissions are being released into the environment from the traditional power stations.

4.12.3. Tax Credits

Not only the individual gets to save money from his or her utility bills, one will also get tax credits and tax benefits from the government and local and state taxes in the context of installing the solar energy systems (Figure 4.8).

Figure 4.8. Benefits of using solar photovoltaic.

Source: Image by pxhere.

4.12.4. Clean Energy

Production of solar energy is a clean energy source. An individual is able to power his or her entire house with the help of the sunlight or solar radiation, and in the grand scheme of things, it is very easy in order to install and maintain the solar energy system.

4.13. ADVANTAGES AND DISADVANTAGES OF SOLAR PHOTOVOLTAIC

The panels of the solar PV system are certainly what come to the minds of people when they talk about the solar energy. Well thought out that in an hour, the sun radiates enough amount of solar energy which can be used in order to cover the human energy consumption for near about a year, then going green with the solar PV panels is maybe in the right direction.

Never the less, with the help of the technology of the solar system, humans are still behind in capturing this naturally free vast amount of the energy that is provided by the nature.

4.13.1. Advantages of Solar Photovoltaic

1. PV panels provide clean as well as green energy. Through the course of the electricity generation with the help of PV panels there is no emission of any kind of harmful greenhouse gas (GHG), and in this way the solar PV is environmentally friendly.
2. Solar energy is the energy, which is provided by nature, and in this way it is free and abundant.
3. Solar energy can be made available almost in any place, where there is presence of the solar radiation or sunlight.
4. More particularly, the solar energy is appropriate with respect to the smart energy networks with distributed power generation (DPG) and it is certainly the next generation power network structure.
5. At the present interval of time, the cost of the solar panels is on a fast reducing track and it is anticipated to continue reducing for the interval of several coming years. As an outcome, the solar PV panels have certainly a highly promising future with respect to both, for economic viability as well as the environmental sustainability.

4.13.2. Disadvantages of Solar Photovoltaic

1. All in all, with respect to the renewable energy sources, the solar energy has intermittency issues, that is not shining at the night time but also, through the course of daytime, when there is a cloudy or maybe rainy weather.
2. As an outcome, intermittency, and randomness of the solar energy makes solar energy panels less reliable a solution.
3. Solar energy panels necessitate additional equipment (it could be inverters) in order to transform the direct electricity (DC) in the form of alternating electricity (AC) in order to be used on the power network.
4. For a continuous supply of electric power, more particularly with respect to the on-grid connections. The PV panels necessitate not only the inverters but also the storage batteries. In this way, increment in the investment cost for PV panels in a considerable amount.
5. In case of land-mounted PV panel installations, the land-mounted PV panel necessitate comparatively large areas for deployment, generally the land space is committed with respect to this purpose for a period of near bout fifteen years to twenty years or maybe even longer interval of time.

4.14 CONCLUSION

In the conclusion of the chapter, the technology of the Fuel cells is quite promising. Fuel cells are having the potential in order to play a significant role as the economy weans itself away from the fossil fuels. The technical difficulties that come in the way can be sorted out with the help of more research and development.

Never the less, the fuel cells have been in to use for a long interval of time. Fuel cells have had less investment when they are compared with the batteries or the internal combustion machines, and are much less mature technology. In simple words, the fuel cells will likely continue to enhance.

Several numbers of the manufacturers are already decreasing the quantity of platinum that is required in their Proton-exchange membrane systems or eliminating it altogether. Manufacturers are also making fuel cell support systems more cost-effective, more reliable, as well as more compact in size.

REFERENCES

1. Acciona, (n.d.) Photovoltaic Solar Energy and Its Contribution | ACCIONA. [Online] Acciona.com. Available at: https://www.acciona.com/renewable-energy/solar-energy/photovoltaic/ (accessed on 4 July 2020).
2. Andreas, L., (n.d.) Solar PV | Student Energy. [Online] Studentenergy.org. Available at: https://www.studentenergy.org/topics/solar-pv (accessed on 4 July 2020).
3. Bethel, A., Microgeneration, A., Jordan, H., Braden, H., James, J., Kailyn, S., Brodie, Y., & Jason, D., (2018). Photovoltaic System-Energy Education. [Online] Energyeducation.ca. Available at: https://energyeducation.ca/encyclopedia/Photovoltaic_system (accessed on 4 July 2020).
4. Brooke, S., (2020). Fuel Cell | Definition, Types, Applications, and Facts. [Online] Encyclopedia Britannica. Available at: https://www.britannica.com/technology/fuel-cell (accessed on 4 July 2020).
5. Brooke, S., (2020). Fuel Cell-Development of Fuel Cells. [Online] Encyclopedia Britannica. Available at: https://www.britannica.com/technology/fuel-cell/Development-of-fuel-cells (accessed on 4 July 2020).
6. Brooke, S., (2020). Fuel Cell-Types of Fuel Cells. [Online] Encyclopedia Britannica. Available at: https://www.britannica.com/technology/fuel-cell/Types-of-fuel-cells (accessed on 4 July 2020).
7. Brooke, S., (2020). Molten Carbonate Fuel Cell | Device. [Online] Encyclopedia Britannica. Available at: https://www.britannica.com/technology/molten-carbonate-fuel-cell (accessed on 4 July 2020).
8. Brooke, S., (2020). Solid Oxide Fuel Cell | Device. [Online] Encyclopedia Britannica. Available at: https://www.britannica.com/technology/solid-oxide-fuel-cell (accessed on 4 July 2020).
9. Collins, J., Gourdin, G., & Qu, D., (2018). Modern applications of green chemistry. Green Chemistry, [Online] pp. 771–860. Available at: https://www.sciencedirect.com/science/article/pii/B9780128092705000285 (accessed on 4 July 2020).
10. Earnest, G., (2020). Solid Oxide Fuel Cells. [Online] Mypages.iit.edu. Available at: https://mypages.iit.edu/~smart/garrear/fuelcells.htm (accessed on 4 July 2020).
11. Evergreen Solar, (2020). How Does Solar Photovoltaic Energy Work?

[Online] Evergreensolar.com. Available at: https://evergreensolar.com/how/how-does-solar-photovoltaic-energy-work/ (accessed on 4 July 2020).

12. Fuel Cell & Hydrogen Energy Association, (2020). Fuel Cell Basics—Fuel Cell & Hydrogen Energy Association. [Online] Fuel Cell & Hydrogen Energy Association. Available at: http://www.fchea.org/fuelcells (accessed on 4 July 2020).
13. Fuel Cell Store, (2020). Solid Oxide Fuel Cell Components. [Online] Fuelcellstore.com. Available at: https://www.fuelcellstore.com/fuel-cell-components/sofc-materials (accessed on 4 July 2020).
14. Renewable Energy World, (2012). Advantages and Disadvantages of Solar Photovoltaic-Quick Pros and Cons of Solar PV-Renewable Energy World. [Online] Renewable Energy World. Available at: https://www.renewableenergyworld.com/2012/12/19/advantages-and-disadvantages-of-solar-photovoltaic-quick-pros-and-cons-of-solar-pv/#gref (accessed on 4 July 2020).
15. Science direct, (2020). Molten Carbonate Fuel Cells: An Overview | Sciencedirect Topics. [Online] Sciencedirect.com. Available at: https://www.sciencedirect.com/topics/earth-and-planetary-sciences/molten-carbonate-fuel-cells (accessed on 4 July 2020).
16. The National Renewable Energy Laboratory, (n.d.) Solar Photovoltaic Technology Basics. [Online] Nrel.gov. Available at: https://www.nrel.gov/research/re-photovoltaics.html (accessed on 4 July 2020).

Chapter 5

Biomass and Nuclear Energy

CONTENTS

This chapter discusses about the two sources of energy which could be very useful for the human beings in the future. Firstly, this chapter explains the fundamental idea of biomass. After explaining the types of biomass fuels, this chapter explains the commonly used methods of extracting the energy from the biomass.

Then, it discusses the major advantages and disadvantages of the extracting and using the biomass energy. In the later phase, this chapter explains the use of nuclear energy. This chapter explains about the functioning of the nuclear reactor and the significance of various elements in the nuclear reactions. Also, the major issue of handling the nuclear waste is discussed in the chapter. Lastly, this chapter explains the advantages and disadvantages of using the nuclear energy.

5.1. INTRODUCTION

Biomass is defined as any kind of organic matter that is crops, seaweed, wood, waste of animal that can be used as the source of energy. After the sun, biomass is likely to be the oldest source of energy. It has been observed that for thousands of years, individuals have burned wood in order to heat their places where they live and cook food for themselves (Figure 5.1).

Figure 5.1. Biomass can be present in any form around the human beings.

Source: Image by Pixabay.

The energy is attained by Biomass from the sun. There is the stored energy, from the sun, in all the organic matter. Sunlight provides the plants with the energy that they require in order to convert water as well as CO_2 into oxygen and sugars. This is done by the process which is known as photosynthesis. These sugars, which are known as carbohydrates, supply the animals and plants that feed on plants with energy.

Foods that are rich in carbohydrates are considered as a good source of energy for the body of the human beings. Biomass is considered as a renewable energy source. This is because biomass supplies are not limited. One can always plant more crops and trees, and there will always be the existence of waste.

Nuclear technology makes use of the energy that is emitted by dividing the atoms of specific elements. Firstly, it was developed in the 1940s, and at the time of the Second World War research initially paid attention on the production of bombs. The focus was shifted, in the 1950s, to the peaceful utilization of nuclear fission, controlling it for the purpose of power generation.

In fact, it has been observed that through regional transmission grids, there are many more countries that are dependent in part on nuclear-generated power. For instance, Italy, and Denmark get near about 10% of their electricity from the nuclear power that is imported.

The commercial nuclear industry was started in the 1960s. When this nuclear industry was started, there were clear boundaries between the industries of the West and East. It has been observed that in the present times, there exist no separate American spheres and Soviet spheres, and this nuclear industry is described by international commerce.

For example, a reactor that is under construction in Asia may have some elements that are supplied to them from Canada, Japan, Russia, South Korea, France, Russia, and some other nations. In a similar way, uranium from Namibia or Australia may end up in a reactor in the UAE. It has been converted in France, and enriched in the Netherlands, deconverted in the United Kingdom, and fabricated in South Korea.

It has to be noted that all the applications of nuclear technology expanded well beyond the provision of low-carbon energy. It helps in controlling the spread of illness and it also helps the doctors in their diagnosis as well as treatment of patients. In addition to this, it powers the most ambitious missions to discover the space. These different uses of nuclear technology positions are at the heart of the world's efforts to attain sustainable development.

5.2. WHAT IS BIOMASS?

Biomass is described as the plant material that is derived from the reaction that took place between Carbon dioxide (CO_2) in the air, water, and sunlight, by the process of photosynthesis, in order to produce carbohydrates. These carbohydrates help in forming the building blocks of biomass.

Generally, the process of photosynthesis converts less than 1% of the sunlight that is available, to stored, chemical energy.

The solar energy that drives the process of photosynthesis is kept in the chemical bonds of the structural components of biomass.

If biomass is processed in an efficient manner, either it is processed chemically or biologically, by extracting the energy that is stored in the chemical bonds and the subsequent 'energy' product that are combined with oxygen, the carbon is oxidized in order to produce CO_2 and water. This entire process is cyclical in nature, as the CO_2 is then made available for the production of new biomass.

The value of a specific kind of biomass is dependent on the chemical properties as well as physical properties of those large molecules from which it is formed. Man, in the modern ages has exploited the energy that is stored in these chemical bonds.

This is done by burning biomass as a fuel and by consuming plants for the nutritional content of their starch and sugar.

It has been observed that in the recent time, the biomass, that has been converted and stored as fossils, has been overused and exploited in the form of coal and oil. Although, it takes a long period of time that is millions of years to transform biomass into fossil fuels, these are not renewable in a time-scale mankind can make use of.

The burning of fossil fuels make use of "old" biomass and then converts the old biomass into "(CO_2)."

This contributes to the effect that is known as greenhouse effect and a non-renewable resource is depleted. It has to be noted that no new CO_2 is contributed to the atmosphere by the burning of new biomass.

This is because replanting harvested biomass makes sure that CO_2 is absorbed and it is returned back for a cycle of new growth (Figure 5.2).

Figure 5.2 Fossil fuels have been used as the main energy source for last several decades.

Source: Image by Flickr.

There is one essential factor that is often ignored at the time of considering the utilization of biomass to assist alleviate global warming, is the time lag that is between the instant release of CO_2 from the process of burning fossil fuels and its final uptake as biomass, which can take several numbers of years.

There are many problems that are faced by the developed countries. One of the problems that faced by the developed world is the necessity to identify this time delay and take some appropriate action in order to mitigate against the lag period. A similar kind of dilemma faces the developing world as it uses its biomass resources for fuel, but it does not execute a program of replacement planting.

It has been observed that there are various crops that have been proposed or these crops are being tested for the purpose of commercial energy farming. Potential energy crops consist of woody crops and grasses plants or herbaceous plants (that is all perennial crops), sugar crops and starch crops and oilseeds. In general, the ideal energy crop has the following characteristics:

- Crops have low cost,
- Have low energy input to produce,
- Have low nutrient requirements,
- Composition with the least contaminants, and
- High yield (maximum production of dry matter per hectare).

5.3. TYPES OF BIOMASS

In the present times, various types of biomass are used. This biomass consists of agricultural products, solid waste, biogas, wood, landfill gas, and biofuels.

5.3.1. Wood

Home grown energy is the biomass that is mostly used in the present times. Wood, that includes chips, bark, logs, and sawdust, accounts for 42% of biomass energy. But it has to be noted that any of the organic matter can produce the biomass energy (Figure 5.3).

Figure 5.3. Wood is one of the category of biomass.

Source: Image by LibreShot.

There are some other sources of biomass. These consist of agricultural waste products such as corncobs and fruit pits. Wood and the waste of wood are used in order to produce electricity. More quantity of electricity is used by those types of industries that make waste. The energy is not supplied by utilities it is a process which is known as cogeneration. In addition to it, paper mills as well as saw mills make most of their waste products in order to generate steam and electricity for their own use.

Although, since they make use of so much energy, they are required to purchase additional electricity from utilities. Gradually, timber companies and those industries that are involved with wood products are seeing the advantages of making use of their lumber scrap and sawdust for the purpose of power generation.

It has many benefits. This helps in saving the disposal costs and, also it has been observed that in some of the areas, it may also reduce the utility bills of the companies. Indeed, the pulp industries and the paper industries are dependent on biomass for well over half of their needs of energy. Some other industries that make use of biomass consist of industries that manufacture furniture, lumber producers, agricultural businesses such as rice and nut growers, and companies that produce liquor.

5.3.2. Solid Waste

Trash turns waste is burned in order to produce a usable form of energy. It has to be noted that one ton (that is 2,000 pounds) of wastes includes about as much heat energy as 500 pounds of coal. Waste is not all biomass. It has been speculated that almost half of the energy content for the biomass comes from plastics, the plastics that are made from natural gas and petroleum. Power plants that burn the waste or garbage for energy are known as waste-to-energy plants.The electricity is generated by these plants just as coal-fired plants do, except for that combustible garbage or waste—not coal—is the fuel that is used to fire the boilers. The generation of electricity from the waste or garbage costs more as compared to the generation of electricity from coal and other sources of energy. Primary benefit of burning solid waste is that it helps in decreasing the volume of garbage that is dumped in landfills by up to 90%. As a result of this, the cost of landfill disposal is also reduced. In addition to this, it also utilizes the energy in the waste or garbage, instead of burying it in a landfill, where it stays unused.

5.3.3. Landfill Gas and Biogas

It has to be kept in our mind that the fungi and bacteria are not picky eaters. They consume dead plants and animals. Thus, causing them (bacteria or fungi) to decay or rot. A fungus that is on a rotting log is converting cellulose to sugars in order to make food for or feed itself.

However, it has been observed that this process is slowed in a landfill, a substance which is known as CH_4 gas is still generated as the garbage or waste decomposes. Regulations need landfills to collect CH_4 gas for the purpose of safety as well as environmental reasons. There are some characteristics of CH_4 gas. This gas is odorless that is it is fragrance-free, and also it is colorless, but it is not safe. That means CH_4 gas is harmful. If this gas seeps into the nearby houses and exploded, it can cause fires or explosions.

The CH_4 gas can be collected by landfills. In landfills, this gas is purified and is used as fuel for the generation of electricity. CH_4 is the most important ingredient in natural gas, and it is a good source of energy. It has been observed that majority of the gas stoves and furnaces make use of CH_4 that is supplied by utility companies. In the year 2003, East Kentucky Power Cooperative started the process of recovering CH_4 from three landfills.

In the present times, the utility make use of the gas at six landfills in order to generate sufficient amount of electricity to power nearly 9,000 Kentucky homes. Still, it has been observed that in the present times, a small quantity of landfill gas is used in order to provide energy. Most of the gas is burned off at the landfill. With the low prices of natural gas, these higher prices of biogas are less economical to collect. However, CH_4 gas is a more powerful greenhouse gas (GHG) as compared to that of CO_2.

It is best thing for the environment to burn landfill CH_4 and convert it into CO_2 by the process of combustion rather than releasing the gas into the atmosphere. The production of CH_4 can also be done by the use of agricultural waste and human wastes. Biogas digesters are defined as the airtight containers or these are the pits that are lined with bricks or steel.

Garbage or waste that is kept in the containers is fermented in the absence of oxygen for the production of a CH_4-rich gas. This produced gas can be used in order to generate electricity, or it is used for various other purposes such as lighting and cooking.

This gas is considered to be harmless and clean-burning gas. This gas produces little carbon monoxide (CO) and it does not produce smoke. Biogas digesters are economical. These are inexpensive to make and maintain. Biogas digesters can be built as community-sized units or family-sized units.

Moderate temperatures as well as moisture are required by them for the process of fermentation to take place. It has been observed that for the developing countries, biogas digesters can the best answers to many of the needs of their energy.

They can also help in reversing the rampant deforestation that is caused by wood-burning, decrease the level of pollution in the air, fertilize the fields that are over-used, and in addition to it, it helps in producing clean, safe energy for the population of rural regions. The use of Biomass Until the middle of 1800s, wood provided the developed nations such as Americans with 90% of the energy that is used in the country. In the year 2016, biomass helped in providing 4.89% of the total energy that was consumed. Biomass has been largely substituted by natural gas, coal, and petroleum.

It has been observed that in the present times, a little less than 42% of the biomass that is used comes from the process of burning wood and wood scraps like saw dust. In addition to it, near about 48% comes from biofuels, mainly ethanol. These are used as a gasoline additive.

The rest of the biomass comes from wastes, crops, and landfill gas. The bigger user of biomass is industry. It has to be noted that more than 48% of biomass is used by the industries. All the electric utilities make use of almost 11% of biomass for the purpose of power generation.

In turn, 1.55% of the electricity is produced by biomass that is used by the people. The other big use of biomass is transportation. Transportation sector make use of about 30% of biomass in order to produce biofuels such as biodiesel and ethanol.

In addition to it, nearly 8% of the supply of biomass is used by the residential sector. There are many studies that showed about 3% of homes in the developed nations such as America make use of wood as the only available source of heat. Also, majority of the houses in the developed nations burn wood in fireplaces and wood stoves for keeping themselves warm.

5.4. METHODS OF EXTRACTING BIOMASS ENERGY

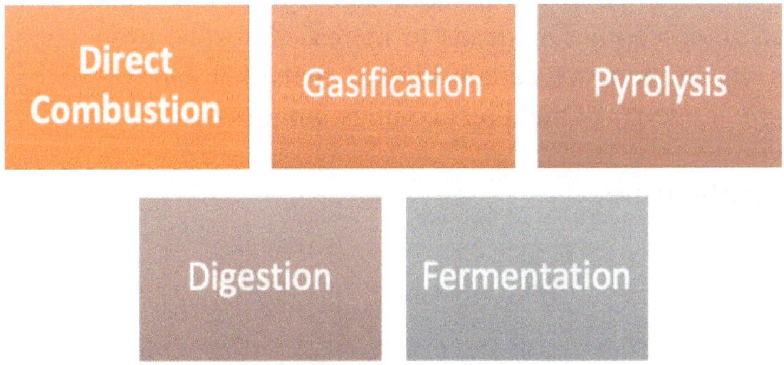

Figure 5.4. Different methods that can be used to extract the biomass energy.

It is possible to convert biomass to thermal energy, liquid, fumes of gas fuels, and other kinds of chemical items. This conversion can take place with the help of number of transformations. It has been observed that the biopower technologies have demonstrated a number of alternatives in the United States, with an installed capacity of 10GW (Figure 5.4).

The entire capacity of producing energy through Biomass totally depends on using the technology of direct combustion. It has been predicted that in the near future, the improvement in efficiency of energy productions by biomass will majorly include the co-firing of biomass in existing boilers which are fired up by the coals.Also, there will be introduction of various high-efficiency processes such as combined-cycle systems, fuel cell systems, and various modular arrangements. Usually, it has been observed that some of the major biopower technologies consist of direct combustion, co-firing, gasification, pyrolysis, anaerobic digestion, and fermentation.

5.4.1. Direct Combustion

This is one of the easiest methods that can be used for extracting energy from biomass. It has been observed that the biomass combustion facilities in the industrial setups are capable of burning different kinds of biomass fuel. Some of the major kinds of fuels that can be combusted in the industrial biomass facilities are wood, residues from agricultural lands, wood pulping liquor, municipal solid waste (MSW), and fuel derived from the refuse collected. After the fuel is collected, the biomass is burned in order to produce steam. The steam produced, at this stage, turns a turbine and the turbine drives a generator. This entire process leads to the production of the electricity. The only disadvantage in the direct combustion process is the ash build-up. Because of the ash build-up, the boilers get fouled and their efficiency gets decreased significantly.This leads to increased cost of production. This is the only reason there are certain kind of fuels that can be utilized for the production of energy in the direct combustion process.

5.4.2. Gasification

Figure 5.5. The entire process of gasification takes place in the gasification plant.

Source: Image by Wikimedia Commons.

Gasification is a process of energy production that uses the high temperature and limited oxygen conditions on solid fuels in order to produce a fuel in its gaseous form. The gas, that is produced, is a mixture of various gases such as CO, CO_2, nitrogen, hydrogen, and CH_4 (Figure 5.5).

The produced gas is used to drive a combined cycle gas turbine that supposedly has a high efficiency. There are various advantages of using the process of gasification over the typical solid fuel burning processes. One of the prominent advantages is having a level of comfort. One of the gases produced in this process is CH_4. CH_4 can be used in place of natural gas under some conditions.

Another major advantage of gasification is that it is capable of producing a fuel that is already free form contaminants and impurities. This leads to less pollution. If conducted under suitable conditions, it can prove to be very helpful in producing synthesis gas. This gas is a mixture of CO and hydrogen.

The synthesis gas can be used to produce Hydro-carbon (such as methane and methanol) for replacing the fossil fuels. In fact, hydrogen itself is a potential fuel which does contain many pollutants and contaminants. Also, it is predicted that hydrogen is capable of replacing oil and petroleum as fuel in the near future.

5.4.3. Pyrolysis

In simple words, pyrolysis is the process that represents the concept of heating the biomass in order to remove the pollutants and contaminants and leaves only the charcoal as residue. This process doubles up the energy density of the original material. This is so because charcoal, which comprises nearly half of the weight of the biomass, contains same amount of energy.

This makes the fuel more transportable. Also, the charcoal requires a much higher temperature for burning as compared to the original biomass. This makes charcoal an increasingly valuable fuel type for various industrial types. There are some sophisticated techniques of pyrolysis. These processes have been recently developed just to collect the volatile particles which are lost in various other processes.The volatile compounds, that are collected while performing pyrolysis, are used to produce a gas rich in hydrogen and CO. It has already been observed that hydrogen is a potential fuel and can be used in a number of various other processes. These gaseous compounds are then converted into a number of chemical compounds such as CH_4, methanol, and a number of hydrocarbons.

Flash pyrolysis is a process that is used to produce bio-crude which is a combustible fuel. Heat produced is then used to perform chemical conversion of biomass into pyrolysis oil. The pyrolysis oil, which is easier to store and transport as compared to the solid biomass material, is then burned like petroleum to produce energy. Also, the process of pyrolysis can be used to convert the biomass into phenol oil. Phenol oil is a kind of substance that is used to make wood glues, molded plastics, and foam insulations.

5.4.4. Digestion

The process of biomass digestion functions by putting the anaerobic bacteria work. These anaerobic bacteria generally live at the bottom of the swampy areas or in several other places where air is absent. These bacteria majorly feed on dead organic matter and produce deliver CH_4 and hydrogen. These anerobic bacteria are used for performing the process of digestion.

For example, the transmitted gas is produced by including microorganisms in natural waste material such as animal excreta or human waste material in the tanks known as the digesters. Later on, the transmitted gas is then collected in order to be used as an energy source.

It has been observed that this process of producing energy is a very efficient process. It enables to extract the maximum of the usable energy from the biomass. Generally, almost 67% of the fuel energy, that is present in the animal dung, can be recovered. Also, this technique is used for collecting CH_4 gas from the landfill sites.

At the landfill sites, huge amount of household biomass waste is dumped. For example, kitchen wastes, organic waste material collected after clipping and pruning the lawns ends up at the local tip. After several decades, the waste matter undergoes the process of steady decomposition. This steady decomposition is carried out by the anaerobic bacteria which are deeply rooted at the bottom of the local tips. These anaerobic bacteria steadily decompose the organic matter and emit CH_4.

The CH_4 gas, which is produced after the decomposition, can be collected and put to use by covering the landfill site with an impermeable layer of clay. In this impermeable layer, perforated pipes are inserted. The gas collected under the impermeable covering is then brought to the surface with the help of these perforated pipes.

5.5. ADVANTAGES OF BIOMASS ENERGY

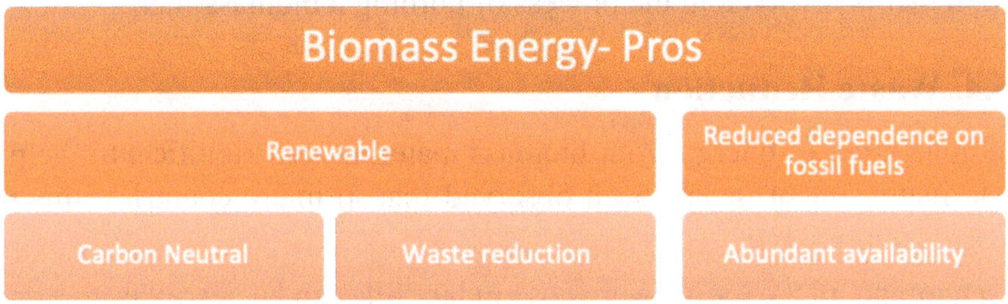

Figure 5.6. Various advantages of biomass energy.

5.5.1. Renewable

As the availability of the sources of biomass like manure, garbage, and dead plants may not diminish as compared to that of the finite fossil fuel, the alternate source of energy is considered to be renewable in nature (Figure 5.6).

Though the use of some of the sources of biomass such as crops may take some time to renew them, depletion of such sources is absent-in some of the cases, renewing of the sources of a biomass can be completed in a few months.

5.5.2. Reduced Dependence on Fossil Fuels

As the sources of biomass can be transformed to fuels and electricity, they can help in decreasing the reliance on fossil fuels.

It has been observed that finite reserves are being a cause of issue in case of fossil fuels. The availability of biomass sources in abundant provide potential to change the energy mix of the world.

5.5.3. Carbon Neutral

It has to be noted that one of the main advantages of biomass energy is Carbon-neutrality. During the process of photosynthesis, plants absorb carbon that is available in the atmosphere and again, this carbon is released back when these plants are burnt in a biomass energy plant.

The new plants again absorb the released carbon, balancing the emissions of carbon that is involved in the process of burning a biomass fuel.

5.5.4. Waste Reduction

The generation of energy from biomass materials can significantly help in waste management. It has been observed that in many countries, this has become a main issue.

Dumping all the garbage or waste in landfills can be decreases to a great extent by growing the number of biomass energy plants.

In addition to it, biomass energy can also have a positive influence on the environment by helping in the reduction of the waste.

5.5.5. Abundant Availability

It has been estimated that biomass energy has a huge potential because of the sources of biomass are available in abundance.

There has been observed an increasing availability of animal waste as well as plant waste, as there continues to be a rise in the level of production and consumption.

5.6. DISADVANTAGES OF BIOMASS ENERGY

5.6.1. Carbon Storage

Carbon is present in biomass. CO_2 is released on combustion and the time to recapture the carbon that is involved in biomass may differ. It depends on the kind of biomass.For instance, the time that is involved in recapturing carbon which is stored in biomass collected from forests might be higher as compared to the other forms.

5.6.2. Environmental Damage

The use of fuels that are produced from biomass results in the emissions of various gases like CO_2, CO, volatile organic compounds (VOCs), nitrogen oxides (NOx). All these gases contribute to air pollution.

It has been observed that in some of the cases, they are found to emit higher particulate matter and other pollutants into the environment as compared to those fuels that are produced by traditional fuel sources.

In addition to it, the usage of forest-based biomass has been criticized by various environmental groups because of its destructive effect on the environment.

5.6.3. Expensive

It has been observed that the construction cost as well as the operating costs of a biomass energy plant can be costly as compared to the traditional forms of power generation.

There is a need of huge space for storage facilities as harnessing energy from biomass includes a selection of various processes.

In addition to it, the process of extraction of biomass materials can be again a costly task because there is the great variability in the sources.

5.7. NUCLEAR ENERGY: THE CONTROVERSIAL FORM OF ENERGY

Nuclear energy is defined as the energy that is in the nucleus, or core of an atom. Atoms are described as the tiny units; these units make up all matter in the universe. The thing that holds the nucleus together is the energy. There is a huge quantity of energy in the dense nucleus of an atom. Officially, the power that holds the nucleus together is known as the "strong force" (Figure 5.7).

Figure 5.7. Nuclear power plant is considered as the powerhouse for the future of the mankind.

Source: Image by LibreShot.

In order to produce electricity, nuclear energy can be used. But first, the nuclear energy must be released from the atom. Atoms are split to release that energy in the process of nuclear fission.

A nuclear reactor, or power plant, is defined as a series of machines that can have control on the nuclear fission for the production of electricity. The fuel that is used by the nuclear reactors for the production of nuclear fission is pellets of the element uranium.

It has been observed that the atoms of uranium are forced to break apart in a nuclear reactor. As the atoms of uranium split, the tiny particles are released by the atoms. These tiny particles are known as fission products. Fission products cause the other atoms of uranium to split thus, beginning a chain reaction. The energy that is emitted from this chain reaction produces heat.

The heat that is produced by nuclear fission warms the cooling agent of the reactor. Usually, a cooling agent is water, but there are some nuclear reactors that make use of molten salt or liquid metal.

The cooling agent of the reactor that is heated by nuclear fission makes steam. The wheels are turned by a flowing current or the steam turns the turbine. The turbines are responsible for driving generators. This can be simply explained with the examples of the engines that are used to produce electricity.

Rods of material known as the nuclear poison are used. These are primarily used in order to regulate the amount of electricity that is produced. Nuclear poisons are defined as the materials, like a kind of the element xenon, that are responsible for absorbing some of the fission products that are made by nuclear fission.

The reaction will be slower, and it will be more controlled if more rods of the nuclear poison are present during the chain reaction. The removal of the rods will permit a stronger chain reaction and it will even produce more electricity.

As of the year 2011, it has been observed that near about 15% of the electricity of the world is produced by nuclear power plants. For example, there are more than 100 reactors that are present in the United States, although it produces most of its electricity from hydroelectric energy and fossil fuels. There are some countries like France, Lithuania, and Slovakia produce almost all of their electricity by the help of nuclear power plants.

5.7.1. Nuclear Food: Uranium

Uranium is defined as the fuel that is most widely used in order to produce nuclear energy. This is because the atoms of uranium get split apart easily as compared to atoms of other radioactive element. In addition to it, uranium is also a quite common element that is found in rocks all around the world. Although, the particular kind of uranium that is used for the production of nuclear energy is known as U-235 and it is rare. In the entire world, U-235 makes up less than 1% of the uranium.

Even though it has to be noted that some of the uranium that is used by the United States is mined in this country but most of the quantity is imported. The United States gets uranium from countries such as Canada, Russia, Australia, Kazakhstan, and Uzbekistan. Uranium must be extracted from other minerals once it is mined. Before Uranium can be used, it must also be processed.

Only those nations that are part of the Nuclear Non-Proliferation Treaty (NPT) can import uranium or plutonium, the other nuclear fuel. This is because nuclear fuel can be used to make nuclear weapons and the nuclear reactors. This treaty encourages the peaceful utilization of nuclear fuel, and it also limits the spread of nuclear weapons.

Near about 200 tons of uranium is used every year by a nuclear reactor. There are some complicated processes that permit some uranium as well as plutonium to be re-enriched or reused. This helps in reducing the amount of extracting, mining as well as processing that is required to be done.

5.8. FUNCTIONING OF THE NUCLEAR REACTOR

Basically, nuclear reactors are large kettles that are used to heat water in order to produce huge amounts of low-carbon electricity. They come in various shapes and sizes and, can be powered by a range of different kind of fuels.

By the splitting of atoms, a nuclear reactor is driven. This process is known as fission. Here, a particle (that is a 'neutron') is fired at an atom, this particle then fissions into two atoms of smaller sizes and some more neutrons. Then, there are some of the neutrons that are emitted, these hit other atoms, causing the atoms to fission too and release some more neutrons. This is known as a chain reaction. In the chain reaction, the process of fission of atoms also emits a huge quantity of energy as heat.

The heat, that is produced, is removed from the reactor by the help of a circulating fluid and generally this circulating fluid is water. Then, this heat can be used in order to generate steam. This drives turbines for the production of electricity. In order to make sure that the nuclear reaction happens at the right speed, reactors include the systems that accelerate, slow or these systems shut down the nuclear reaction, and the heat that it produces.

Generally, this is carried out with control rods. Control rods are generally made of neutron-absorbing materials like born and silver. There are many different shapes and sizes of nuclear reactors. Some of these nuclear reactors make use of water to cool their cores. On the other hand, some other nuclear reactor makes use of gas or liquid metal. It has been observed that the most common kinds of power reactor make use of water, with over 90% of the reactors of the world being water-based.

Some of the other information about the various different kind of reactor all over the world can be found in the Nuclear Power Reactors section of the Information Library. It has to be noted that nuclear reactors are very reliable at producing electricity. In addition to it, these nuclear reactors can run for 24 hours a day for several months, if not years, without any kind of interruption, irrespective of any kind of the weather or season.

Moreover, most of the nuclear reactors can functions for very long periods of time. It has been observed that these reactors operate for more than 60 years in many of the cases. In the year 2019, units 3&4 at the Turkey Point plant in Florida were the first reactors in the whole world that were licensed for 80 years of operation.

5.8.1. What Fuels a Nuclear Reactor?

There are various different materials that can be used in order to fuel a reactor, but uranium is the most common material that is used. Uranium is abundant and it can be found in various places all over the world, which includes the oceans. There are various other fuels, like plutonium and thorium that can also be used.

It has been observed that most of today's reactors consist of various hundred fuel assemblies, each of these assemblies are having thousands of small bits of uranium fuel. It has to be noted that a single pellet of uranium fuel contains as much amount of energy as it is available in one ton of coal.

A typical reactor needs near about 27 tons of fresh fuel every single year. In the contrary, a coal power station of the same size would need more than two-and-a-half million tons of coal in order to generate as much electricity.

5.8.2. Radioactive Decay

Nuclear fission can take place either naturally or it can be induced. The process is known as radioactive decay if it takes place naturally and this process is a common process. All the components or elements in which this process takes place naturally are called radioactive isotopes.

These radioactive isotopes are described as those atoms that have an unstable nucleus. This is because there is an excess amount of energy in the nucleus. This instability will cause these atoms to spontaneously decay or these atoms will split into two atoms of the smaller size. These atoms, that are newly formed, will be of different elements or component from the original radioactive isotope and also, the nuclei of these atoms will be more stable.

The prediction of a specific radioactive isotope to decay is not possible, although the average rate at which the radioactive isotope will decay is known; this rate is known as the "half-life." Half-life indicates to the amount of time for a particular quantity of a decaying element to be reduced by half. Half-lives can differ significantly, and it can vary from mere seconds to so many years.

Carbon-14 is one of the examples of naturally occurring radioactive decay. Cabon-14 takes place naturally in the atmosphere, and in the living things it is present at a very certain amount. In order to estimate the age of organic remains, Carbon-14 is used. It is estimated by calculating the quantity of carbon-14 remaining in a fossil and comparing that amount of carbon-14 to the amount that it would have had at the time of death. The half-life of carbon-14 plays an important role while determining the age of the fossil. This can be done by knowing the amount of carbon-14 that has decayed. This decayed amount give a rough estimate of how much time it would have taken for that to happen can be made.

5.8.3. Nuclear Reaction

Another kind of nuclear decay which is known as fission is a manmade nuclear reaction. This reaction takes place inside the nuclear reactor and it produces the heat in order to generate electricity. A nuclear reactor is defined as the steel vessel which contains the nuclear fuel. A nuclear reaction takes place when an atom is caused to split or "fission," and as a result of it, a huge quantity of energy is released.

All those substances that can be induced to split are called as the fissile materials. It has to be noted that it is out of fissile materials that nuclear fuel is made. Uranium is the nuclear fuel that is most commonly used for commercial nuclear power plants. Uranium is denoted by the chemical symbol U. Uranium is defined as a metallic chemical element that is commonly found on the crust of the earth. Uranium that is occurred naturally comes in ten three different types or isotopes.

It has been observed that all the uranium isotopes consist of 92 protons in their nucleus. This is what makes them uranium. The main difference that is between the isotopes is the number of neutrons that are present in the nucleus. Naturally occurring uranium consists of U-234, U-235 and U-238. U-234 is having 142 neutrons.

On the other hand, U-235 is having 143 neutrons and there are 146 neutrons in U-238. It has to be noted that in nature 99.3% of the uranium is U-238, U-235 is of 0.7% and less than .01% is U-234. Among all the naturally occurring uranium, U-235 is the only fissile, naturally occurring uranium isotope. Thus, it is the most used nuclear fuel.

A nuclear reaction begins with a U-235 atom. This atom is induced to split by the process of shooting a neutron at it. When the neutron is absorbed by U-235 atom, it quickly becomes a U-236 atom. It has been observed that the nucleus of the U-236 atom is comparatively unstable. It tends to break up quickly into two different atoms. As it breaks into two different atoms, it releases two to three free neutrons in the process. The atoms, which are formed during the fission process, are known as fission products. The atoms formed as fission products are usually half the size of the original atom.

There will be variations in the kinds of fission products with the progression of each and every nuclear fission reaction. The fission products generally range from elements with atomic number 30 (Zinc), to atomic number 71 (Lutetium). There is a great significance of the free neutrons that are released during the process of fission. These free neutrons are very necessary for the continuation of the nuclear reaction or the nuclear chain reaction.

5.8.4. Nuclear Fuel Cycle

Nuclear power plants make use of uranium as fuel in order to produce electricity. Uranium is considered as a mildly radioactive metal. This metal is very common on the crust of the earth. Uranium must be extensively processed before it can be used as a nuclear fuel. The uranium goes through different phases or stages in the process of producing electricity. These stages are called the nuclear fuel cycle. There are some activities at the front end of the cycle that is associated to the production of nuclear fuel. The service period defined as the period when the nuclear fuel is being used in order to produce electricity. On the other hand, the back end of the cycle indicates to the storage, reprocessing as well as final disposition of the spent nuclear fuel.

5.8.5. Chernobyl

The experts dealing in nuclear energy worry that all the storage facilities for radioactive waste will crack or they will leak or erode. Then, the radioactive material could pollute the soil and groundwater that is near the storage facility. This could give rise to various serious health problems for the individuals as well as for the organisms in the area. All the communities in that particular area are required to leave and evacuate that area.

For example, this situation took place in Chernobyl, Ukraine, in the year 1986. A steam explosion that took place at one of the power plants four nuclear reactors gave rise to a fire, which is known as a plume. This plume was highly radioactive. It created a cloud of radioactive particles. All these radioactive particles fell to the ground, which is known as fallout. The fallout spread all over the Chernobyl facility, and also it spread in the nearby area. This fallout was also drifted with the wind, and the radioactive particles come into the water cycle as the form of rain. This radioactive fallout fell over Scotland and Ireland as rain. It has been observed that in Belarus most of the radioactive fallout fell.It has been observed that over 100,000 individuals were moved after the Chernobyl disaster, but the total number of human victims of Chernobyl disaster is very hard to determine. The impacts of radiation poisoning only occur after several years. Cancers as well as other illnesses can be very hard to trace to a single source.

5.9. HANDLING THE NUCLEAR WASTE

The nuclear industry also generates the garbage or waste in similar manner to any other industry. Although, nuclear power is generated in very small quantities, as compared to the other industries, while managing and containing the amount of energy that is produced. It has been observed that most of the waste from nuclear power plants is not very radioactive and also, for many years, this waste has been correctly managed and disposed of.

It has to be noted that just about 5 grams of highly-radioactive waste would be produced if nuclear power was used to supply the electricity needs of an individual for the whole year. This weigh is the same weight as a sheet of paper. Also, there are several different ways in which the used fuel which comes out of the reactor can be managed. These ways consist of recycling for the purpose of energy production or direct disposal. In fact, there are various nations that have been making use of recycled fuel for years in order to partly fuel their reactors.

5.10. ADVANTAGES OF NUCLEAR ENERGY

In spite of all the potential disadvantages and the controversy that surrounds these shortcomings, nuclear energy also has some benefits over some other ways of energy production. The production is not costly, it is reliable and there is no production of GHGs.

5.10.1. Expense

It was stated by The World Nuclear Association (WNA) that less uranium is required in order to produce the equal amount of energy of oil or coal. This reduces the cost of producing the equal amount of energy. Uranium is inexpensive to obtain and transport. Because of this, it reduces the cost.

In accordance to the Nuclear Energy Institute (NEI), "One uranium fuel pellet creates as much energy as one ton of coal, 149 gallons of oil or 17,000 cubic feet of natural gas."

5.10.2. Reliability

Nuclear power plant can run continuously for one to two years when it is functioning in a proper manner. In accordance to World Nuclear News (WNN) UK's Heysham II plant ran without the need of refueling for a time period of 940 days in the year 2016. As a result of this, there were fewer brownouts or some other power disruptions. The functioning of the nuclear power plant is also not dependent on weather or foreign suppliers. This characteristic of nuclear power plant makes it more stable as compared to the other available types of energy.

5.10.3. No Greenhouse Gases

While there are some harmful emissions produced by the nuclear energy but there are no traces of GHGs being emitted by the nuclear energy power plant. There are many studies that have shown that the life-cycle emissions from the nuclear power plants are more or less similar to the emissions that are given out by renewable energy sources such as the wind power. Several studies were reviewed by WNA and it concluded that "GHG emissions of nuclear power plants are amongst the lowest of any method of electricity generation and on a lifecycle basis are similar to hydro-electricity, wind, and biomass." Also, the low GHG emissions can be very appealing to some of the consumers.

5.11. DISADVANTAGES OF NUCLEAR ENERGY

It has been observed that often the nuclear energy falls under fire. This is because of many of the drawbacks that it brings water pollution, uranium leaks, waste, and reaction failures.

5.11.1. Raw Material

In the process of fission, Uranium is an element that is naturally unstable. This means that there are some special precautions that must be taken during the process of mining of the uranium, transporting, and storing of the uranium. In addition to it, it also includes the process of storing of any waste product in order to prevent it from giving off dangerous levels of radiation. These special precautions are described by the National Academies Press.

5.11.2. Water Pollutant

In accordance to the Department of Physics Stanford University, the water cools the nuclear fission chambers, in both the pressurized water reactors (PWRs) and the boiling water reactors (BWRs). In case of PWRs, there is the production of steam in an indirect manner by running cold water via primary pipes. The heated water is removed away by the secondary pipes. This is done so that the coolant is not in contact with the reactor. On the other hand, in BWRs, the steam is produced in the direct manner as the water runs via the reactor core, so if in case there is any leakage of fuel, the water can get polluted and it is moved to the rest of system.

5.11.3. Spent Nuclear Rods Potential Hazard

It is explained by The United States Nuclear Regulatory Commission (US NRC) that the used nuclear rods are dipped down in water in the spent fuel pool that is below 20 feet of water. These are immersed in order to cool them prior to their transportation for disposal. It has been observed that radioactive water can leak out of doors in the pool when the seals that are used for keeping the doors airtight are not functioning properly. It was observed in the 2013 Japanese Fukushima nuclear plant disaster.

5.11.4. Aquatic Life Dangers and Threats

It was encapsulated by Nuclear Information and Resource Service (NIRS) that how pollutants or contaminants that are emitted by nuclear plants are heavy metals and poisonous pollutants that provide damage to the plant and animal life in aquatic bodies. After being cooled, the water is released into the atmosphere, but the water is still warm, and it provides damages to the ecosystem of the sinks it flows into.

5.11.5. Waste

When the uranium has gone through the process of splitting, the resultant radioactive byproducts are required to be removed.

5.11.6. Leaks

Nuclear reactors are created with various safety systems that are designed to contain the radiations that are given off in the process of fission. When these safety systems are installed as well as maintained in the proper way, they function in an adequate manner. On the other hand, when these safety systems are not maintained in a proper way, have structural flaws, or were installed in an improper way, a nuclear reactor could emit the dangerous amounts of radiation into the environment at the process of regular use.

5.11.7. Shutdown Reactors

It has been observed that there have been various nuclear reactors that have not been successful and been shut down that are still in existence. All these neglected nuclear reactors are taking up valuable land space, could be polluting the areas that are surrounding them, yet these are often too unstable to be eradicated.

5.12. CONCLUSION

Biomass and Nuclear energy are the two major forms of energy that can prove to be very useful for the human beings in the near future. Biomass is the renewable source of energy which is capable of solving the issues of shortage of energy for the human beings. This is majorly because biomass is something which will be produced till the time human beings exist and consume different products. On the other hand, nuclear energy has always been considered as the controversial form of energy. This is so because there is a great amount of risk involved while producing the nuclear energy. Thus, it is very important to understand the pros and cons of both forms of energy and then use any of them wisely to fulfill the energy needs and requirements of the human beings.

REFERENCES

1. (2018). Biomass. [e-book] pp. 10–14. Available at: https://www.need.org/Files/curriculum/infobook/BiomassS.pdf (accessed on 4 July 2020).
2. Asaff, B., (n.d.) Advantages and Disadvantages of Nuclear Energy. [Online] Greenliving.lovetoknow.com. Available at: https://greenliving.lovetoknow.com/Advantages_and_Disadvantages_of_Nuclear_Energy (accessed on 4 July 2020).
3. Giraldo, J., Gotham, D., Nderitu, D., Preckel, P., & Mize, D., (2012). Fundamentals of Nuclear Power. [e-book] State Utility Forecasting Group, pp. 1–76. Available at: https://www.purdue.edu/discoverypark/sufg/docs/publications/SUFG%20nuclear%20report.pdf (accessed on 4 July 2020).
4. Khan, S., Paliwal, V., Pandey, V., & Kumar, V., (2015). Biomass as Renewable Energy. [e-book] Ghaziabad. Available at: https://www.researchgate.net/publication/319456142_Biomass_as_Renewable_Energy (accessed on 4 July 2020).
5. McKendry, P., (2001). Energy Production from Biomass (Part 1): Overview of Biomass. [e-book] pp. 1–10. Available at: https://faculty.washington.edu/stevehar/Biomass-Overview.pdf (accessed on 4 July 2020).
6. National Geographic Society, (n.d.) Nuclear Energy. [Online] Available at: https://www.nationalgeographic.org/encyclopedia/nuclear-energy/ (accessed on 4 July 2020).
7. Nsenergybusiness.com, (2019). What are the Major Pros and Cons of Biomass Energy? [online] Available at: https://www.nsenergybusiness.com/features/newsmajor-pros-and-cons-of-biomass-energy-5845830/ (accessed on 4 July 2020).
8. World-nuclear.org, (2020). Nuclear Power Today | Nuclear Energy-World Nuclear Association. [Online] Available at: https://www.world-nuclear.org/information-library/current-and-future-generation/nuclear-power-in-the-world-today.aspx (accessed on 4 July 2020).
9. World-nuclear.org, (n.d.) How Does a Nuclear Reactor Work-World Nuclear Association. [Online] Available at: https://www.world-nuclear.org/nuclear-essentials/how-does-a-nuclear-reactor-work.aspx (accessed on 4 July 2020).

Chapter 6

Bioenergy and Biofuels

CONTENTS

Biofuels have now become a key part of sustainable living. They have replaced the extensive use of fossil fuels. Being renewable source of energy, they are being used all around the world at large scale. The chapter initially talks about fundamentals of biofuels and bioenergy along with some conversion processes of biofuels.

Later, a brief discussion on sustainability of biofuels is provided. Towards the middle part of chapter, genomics of sustainable bioenergy is explained together with chemistry of certain biofuels. At the end, future of biofuels is discussed in bit detail.

6.1. BIOFUELS AND BIOENERGY BASICS

Different from other renewable sources of energy, biomass can be directly converted into liquid fuels. These liquid fuels are further known as biofuels. They can efficiently aid to meet the requirements of transportation. The two most common kinds of biofuels that are currently being used are biodiesel and ethanol. Both these biofuels represent the first generation of biofuel technology.

The Bioenergy Technologies Office (BETO) is collaborating with different countries in order to develop next-generation biofuels basically made from non-food resources such as cellulosic and algae-based products.

Over the past ten years, BETO emphasized more on cellulosic ethanol and invested largely in technology advancement throughout the supply chain. All such activities significantly validated the technologies for the production of cellulosic ethanol (Figure 6.1).

Figure 6.1. Biofuel is a major part of the recyclable sources of energy.

Source: Image by Wikimedia Commons.

Bioenergy can be described as any energy source which is completely based on the biological matter that is everything from a dung cooking fire or a biomass power station to ethanol-based car fuel.

Different from oil, gas, and coal, bioenergy is considered as a renewable energy as plant as well as animal materials can be regenerated in easy manner. Currently, bioenergy mainly accounts for the production of majority of renewable energy all around the world.

At times, the term biofuels are interchangeably used with bioenergy. However, commonly it is used to describe liquid bioenergy fuels like biodiesel (a substitute of diesel) and bioethanol that can be used in petrol engines.

Often, bioenergy is regarded as eco-friendly as theoretically, the release of carbon dioxide (CO_2) when trees and plants are burned is balanced out by the CO_2 absorbed by the new planted plants in order to replace the harvested ones. Although, the social as well as environmental benefits of bioenergy are hotly contested-mainly in the case of biofuels that are usually produced from the food crops for example corn, palm oil, and sugar.

On the other hand, there are two key criticisms of biofuels. The first one is that growing plants just for energy puts heavy pressure on the use of land, causing deforestation and increasing food prices. The second one is that the formation of certain biofuels has been associated with large carbon emissions.

For instance, producing ethanol from corn can eventually generate high amount of greenhouse gases (GHGs) by the use of fertilizers, tractors, and processing plants. In a similar manner, biodiesel produced from palm oil can have an extreme carbon footprint along with serious negative consequences-if virgin rainforest is cleared so as to make way for the plantations of palm oil.

Promoters of bioenergy argue that though some of the biofuels are unsustainable, while other like biodiesel produced from use cooking oil and burned wood in combine's heat and power plants-make better environmental sense.

It has been observed that the production as well as the consumption of liquid biofuels for transport is currently highly concentrated. The US and Brazil account for 60–70% of world ethanol production. Germany and France account for nearly 60% of biodiesel production and consumption.

FACTS

Biomass accounts for approx. 11% of total primary energy consumed globally and for about 80% of renewable energy, but not all of this is used in a sustainable manner.

Currently, over 85% of biomass energy is consumed as solid fuels for cooking, heating, and lighting, often with low efficiency. Traditional bioenergy (fuelwood, charcoal which can only deliver heat) dominate bioenergy consumption in developing countries where up to 95% of national energy consumption relies on biomass.

Liquid biofuels account for around 2% of road transport fuels worldwide but growth rates and future potential are significant. Current biofuels are bioethanol (based on sugars and starches) and biodiesel (plant oils and animal fats).

In addition to that, they asserted that the next generation biofuels like cellulosic ethanol produced from wood, grasses or straw will eventually create sustainable fuels just by decreasing the requirement of cropland, fertilizers, and processing energy.

A recent criticism on bioenergy is that it would make much sense to use all of the available wood to make furniture and produce building materials, instead of just burning it.

As by doing this, the carbon content gets locked up safely in products and building for several years or even centuries.

Another significant way that can be used to lock up the CO_2 captured by plants and trees would be to burn them in power plants, but the power plants must be equipped with carbon capture as well as storage technology.

Bioenergy will remain important over the coming decade so as to reach the renewable energy targets and that is why member states of the European Union (EU) incorporated the bioenergy option in their National Renewable Energy Action Plans (NREAPs).

As the JRC aims to offer independent scientific advice regarding the use and production of biofuels, it evaluates the environmental sustainability, assesses the technological developments, estimates the use of land because of an increased demand of biofuels, calculates the direct as well as indirect emissions from biofuels, change in land use and distinct bioenergy pathways. The JRC analyses the availability of straw and some other types of biomass for energy purposes.

All the costs and benefits have to be considered to guarantee a well-balanced policy. Assessment of resources is really important. Knowing exactly how much resource is present and at what cost and with which environmental impact, indeed gives a much better understanding of some environmental parameters such as need of water, soil erosion, degradation, and the impact on biodiversity.

While carrying out sustainability and resource evaluations, the considerations extend beyond the borders of EU. As a key part of biofuels is expected to come from some of the tropical nations such as Malaysia, Brazil, Indonesia, the JRC also considers issues that may have a positive or negative impact there: soil and air quality, water footprint, food prices and farm incomes.

The NREAPs state that around 9% of total transport fuels will be attained through biofuels, all of which would come from so-called first-generation biofuels. All these are fuels produced from the crops or used for food that can eventually lead to direct (crops for biofuels grown on uncultivated land) or indirect (crops for biofuels diverted from existing food production) land use change.

In turn, this will result in a certain amount of GHG emissions. The JRC developed a different methodology for the estimation of emissions from direct as well as indirect land use change depending upon the economic models.

To avoid future competition for land use and resources between bioenergy, wood, food productions, agricultural crop residues are expected to be a much relevant source of bioenergy. A spatially explicit assessment of the agricultural crop residues and a straw-availability map has been produced.

6.2. BIOFUEL CONVERSION PROCESSES

6.2.1. Deconstruction

Production of advanced biofuels such as renewable hydrocarbon fuels and cellulosic ethanol usually involves a multistep process. First of all, a tough rigid structure of the plant cell wall that comprises biological molecules cellulose, hemicellulose, and lignin tightly bounded to each other-must be broken down. This can be easily attained in one of the two ways: high-temperature deconstruction or low-temperature deconstruction.

6.2.1.1. High-Temperature Deconstruction

High-temperature deconstruction effectively makes use of extreme pressure and heat in order to break down solid biomass into liquid and gaseous intermediates. There are three major routes that can be used in this pathway:

1. Pyrolysis;
2. Gasification; and
3. Hydrothermal liquefaction.

In the process of pyrolysis, biomass is repeatedly heated at high temperatures of 500°C–700°C in an oxygen-free environment. The heat breaks down the biomass into pyrolysis vapor, char, and gas. Once the char is removed, the vapors are cooled down and then condensed into liquid "bio-crude" oil.

The process of gasification follows a slightly similar process in which biomass is exposed to a high temperature range that is more than 700°C with some oxygen present to produce synthesis gas. Synthesis gas is also known as syngas. Syngas is a mixture that consists mostly of carbon monoxide (CO) and hydrogen.

While working with wet feedstock's with algae, the process of hydrothermal liquefaction is the most preferred thermal process. This process makes use of water under moderate temperature of 200°C–350°C and high pressures to convert biomass into liquid bio-crude oil.

6.2.1.2. Low-Temperature Deconstruction

Deconstruction at low temperature typically uses biological catalysts that are known as enzymes or the chemicals to breakdown the feedstock's into intermediates. First of all, biomass undergoes a step of pre-treatment that opens up the physical structure of plant and algae cell walls, make up sugar polymers such as cellulose and hemicellulose more accessible. Such polymers are further broken down chemically or enzymatically into simple sugar building blocks during a process known as hydrolysis.

6.2.2. Upgrading

Following the process of deconstruction, certain intermediates such as syngas, crude bio-oils, sugars, and other chemical building blocks should be upgraded in order to produce a finished product. This step can comprise either chemical or biological processing (Figure 6.2).

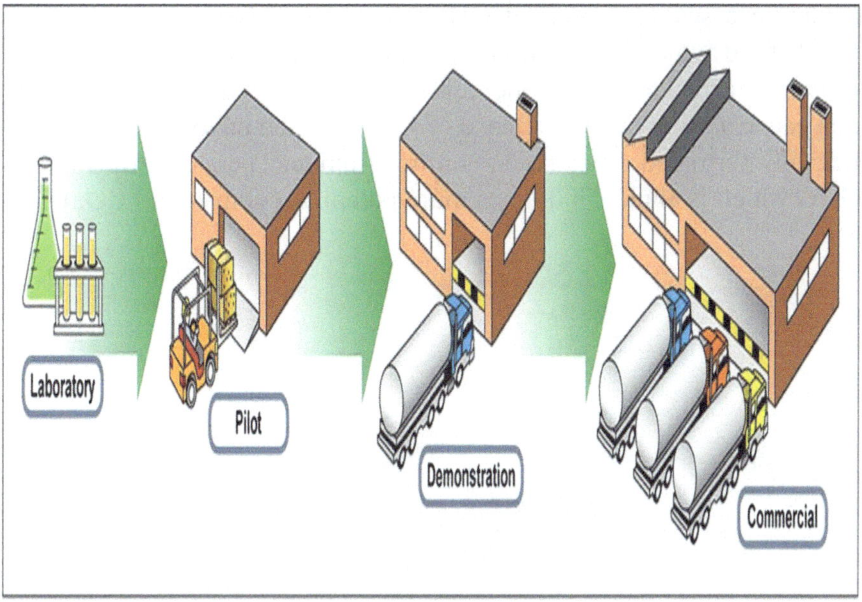

Figure 6.2. Stages of scaling up of the biofuels.

Source: Image of Flickr.

Microorganisms like yeast, bacteria, and cyanobacteria can ferment sugar or gaseous intermediates into fuel blend stocks and chemicals. On the other hand, sugars, and other intermediate streams like syngas and bio-oil may be processed by making use of a catalyst in order to remove any unwanted reactive compounds to improve the storage and handling properties.

The finished products from upgrading may be fuels and bioproducts that are ready to sell into the commercial market or stabilized intermediates that are suitable for finishing in a petroleum refinery or chemical manufacturing plant.

6.3. ARE BIOFUELS SUSTAINABLE?

At present, no biofuel is carbon-neutral. However, GHG emissions from each liter of burned fuels in an internal combustion engine are absorbed by the plants that are grown to produce next liter of fuels, total emissions of CO_2 are released by the fossil fuels used in the manufacturing, agricultural, and distribution processes that produces the fuel.

In extreme cases, emissions of GHGs from fossil fuel inputs can exceed those that are avoided by the user. In addition to that, any grassland or forest that is lost to make way for the cultivation of feedstock oil, starch, sugar crops cause a huge one-off release of CO_2, and the ongoing production of artificially fertilized crops releases nitrous oxide. The gas nitrous oxide is a GHG which is almost 300 times more potent as compared to CO_2 (Figure 6.3).

Figure 6.3. C-17 flight being tested to run on biofuel.

Source: Image by wpafb.

Add in the social cost of high food prices, the environmental impact of habitat loss, water depletion and soil erosion as rigorous biofuel agriculture expand all around the world, and the arguments against continued political support for biofuels in the UK and EU appear to be irresistible.

People should surely wait until better alternatives are available, possible back to the most promising technologies with funding of government and at the same time look at some efficiency measures to limit transport emissions. However, the argument seems persuasive, and is partially based on facts. Although, it misses four important points and hence, draws imperfect conclusions.

Road transport fuels are mainly produced from the oil reserves that are being rapidly depleted at a rate in excess of new discovery. The demand of transport is increasing all around the world, driven by China and India, and scarcity of oil is leading to increase in use of some alternative sources of transport fuels like tar sands, shale, heavy oil, and coal-to-liquid processes (Figure 6.4).

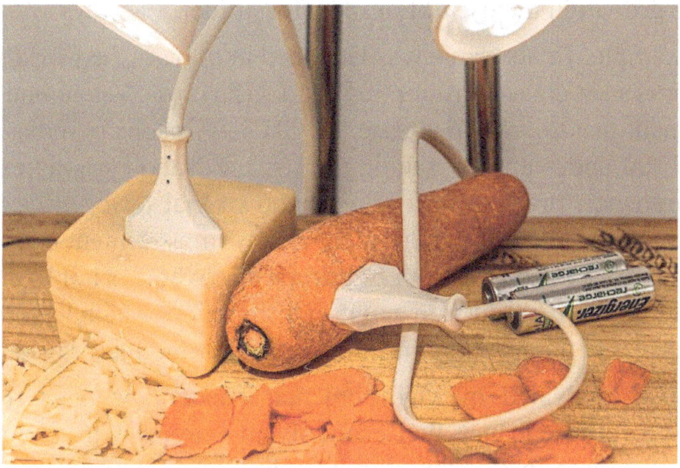

Figure 6.4. Biofuel is being increasingly used to act as alternate sources of energy.

Source: Image by Pixabay.

Being more expensive as compared to the fuel from conventional oil reserves, the sources are environmentally damaging in terms of local effect of extraction and refining as well as in terms of GHG emissions.

For instance, the extraction of oil from tar sands needs large quantities of steam and the fuel hence produced causes at least 50% more GHG emissions as compared with the extraction as well as use of conventional crude oil. Coal-to-liquid process technology is quite less efficient, with around a third of the coal's chemical energy lost as waste heat in the conversion process.

In the future, the extraction and use of remaining conventional oil reserves will produce higher GHG emissions as compared today, owing to the small size and geographic inaccessibility of the remaining productive fields.

This double-whammy of increased demand for damaging fuel creates an imperative for action. A moratorium on EU biofuel targets as suggested by the EAC implicitly endorses these more polluting alternatives. Though, is the answer that makes people drive smaller cars to use less oil? A point that is missed by the EAC concerns the fundamental economics of transport emissions growth. Efficiency is highly desirable from the perspective of a consumer as it decreases the cost of monitoring, though will not decrease the aggregate growth in the use of transport fossil fuel.

For example, Tata's Nano, was launched in India as the world's cheapest car. It claims an average consumption of fuel (50 mpg). Yet, nobody suspects for a moment that this will decrease the GHG emissions in India. At a price of just £1000 and with low running costs, the Nano car was expected to create a transport revolution. It is well known that lower price stimulates the growth of demand.In a similar manner, any improvement in average fuel economy in the United Kingdom or Europe as a whole is not likely to show the global rate of oil production and its consumption, which it would require to do to have a key impact on the emissions all around the world.

Whether motivated by environmental concern or thrift, regulation, ongoing efforts to enhance fuel economy in Europe will be offset by an equivalent increase in consumption anywhere else, until the new demand takes up the slack (stimulated by the temporarily lower oil prices).

The practical approach to assure that fossil fuels are left in the ground is to produce a large number of cheap alternatives. Larry Page, co-founder of Google and champion of Google's renewable energy initiative, recognizes this need: "We want to produce one gigawatt of renewable energy capacity that is cheaper than coal. We are optimistic this can be done in years, not decades."

In a similar way, the aim of a transport biofuels policy should be to produce cheap as well as sustainable alternative to oil as early as possible.

6.3.1. Rising to the Challenge: Industry, Policy, and Technology

The necessary second-generation biofuel technology will be invented in a timely fashion and until this is available, people should not support an industry of biofuel that are based on first-generation technology. When Imperial and Cambridge College bid against three rival universities of US for $500 million of funding by BP to establish an Energy Biosciences Institute, the winner was Berkeley.

Generous state funding played a key part in this decision, but the growing biotechnology and the process technology start-ups across whole North-America developing second-generation biofuel technology, financed by mainly non-government capital, bear witness to the commercial attractiveness of this sector. Investment on this scale depends largely on the availability of first-generation biofuels, which in the US means 'ethanol.' With an annual output of almost 25 billion liters, the US has overtaken Brazil as the largest bioethanol producer of the world and the output is expected to double in the future.

The ethanol industry of the US receives generous state support and is highly based on the corn feedstock by using first-generation technology that is not much better for the environment as burning the oil is displaces. However, this industry has created a climate of investment that has improved the development of second-generation technology.

First-generation ethanol plants offer a much valuable market for the incremental technology in the form of better feedstock varieties microbes, enzymes, and certain other process enhancements. They cause the development of infrastructures such as facilities of storage, distribution networks, and flex-fuel vehicles that are important for the growth of bioethanol consumption in future. They also create the experience as well as know-how that venture capital investors look for in start-up management teams.

The marginal environmental benefits of corn ethanol are reflected by the marginal economics. In the past few years, rising feedbacks and costs of energy have left several of the US plants operating at or below the breakeven.

It is mainly for this reason that advancements in the process technology are valuable for this industry, and why the global center of second-generation bioethanol research and expertise has developed in North America, not in Brazil where first-generation economics are perfectly adequate.

Similar to the US, the UK is a leader of life sciences and process engineering in the whole world. Unlike the US, when it comes to commercialization of capabilities, rest of the world have relatively a poor track record, as it requires clear policy incentives, a solid industrial base, and regulatory frameworks to develop.

The number of start-ups of bioethanol technology today can be counted on the fingers. This is mainly the case until the UK has a credible and established bioethanol production industry with certain commercial requirements and money to spend.

Current biofuel research is basically active in three overlapping areas, and these are feedstock development, process improvement and new process design. In each of these areas, a large number of publics, corporate, and start-up organizations are racing to bring new technologies to market.

In some cases, these are incremental and near term. However, a number of technologies in development have potential to transform the global industry all around the world and anyone of these may develop as the prevailing biofuel process route in the future.

Biodiesel that is made from the fast-growing algae, thermal depolymerization of organic waste to form biocrude, enzyme hydrolysis of forest waste and grasses, and even direct biological synthesis of more complex biofuels each have such potential.Each of them may also be rendered impotent by an insurmountable technical or economic hurdle. Failures in innovations are noted. Only those technologies that offer much better returns on the resources are further adopted.In the market for research finding, various early stage technologies that show promise receive investment, and most die when their potential is shown to be flawed or if a more promising competing alternative shows up.

In contrast, government-funded research tends to be concentrated in those projects where there is political capital to be made.

Once the initial investment sunk, it became really tough for support to be withdrawn without damaging suggestion that funds have been wasted or promise broken. It often ends up backing the wrong technology.

Government research and development in general, accounts for less than 15% of the overall investment in alternative energy.

In Europe, this figure is still lower and the likelihood is that the majority of improvements in the field of biofuels technology will be developed by the private sector to clear policy signals and aligned closely with the needs or requirements of a commercially viable biofuel industry.

The main role of policy is to develop enough up-side potential for the producers as well as technology developers to guarantee a high level of investment required for advancement and innovation towards the most sustainable as well as efficient use of resources.Particularly, the industry needs coherent and clear policy signals that provide a long-term favorable framework for development.

6.4. SUSTAINABLE BIOENERGY: GENOMICS AND BIOFUELS DEVELOPMENT

To address the global energy genomics may be a key factor whether it is believable or not. How to harness various renewable energy sources is one of the important questions. This question can be easily answered by the means of their knowledge in this field.

Also, it is observed that a better understanding is developed by the researches by their knowledge in this particular field. The renewable energy resources which are harnessed are such as lignocelluloses biomass, microalgae, and Cyanobacteria.

Further to optimize development of sustainable biofuels the genetic engineering of enzymes will be surely act as a key factor. Also, someday fossil fuels on a global scale could be easily replaced by these biofuels.

6.4.1. First-Generation versus Second-Generation Biofuels

Biofuel scientists are involving the use of first-generation versus second-generation biofuels and basically, this is one of the primary dilemmas which are currently being faced. Ethanol and biodiesel are primarily consisted in the first-generation biofuels.

These are mainly derived from the oils, sugar, starches. Moreover, it is observed the food crops for the use of agricultural land and water are compete by the crops which are used to these fuels. The environment can be harmed and deforestation, pollution, water use, and release of gases than with fossil fuels (Fargione et al., 2008) and all of this can be caused by the process of formation of these biofuels (Figure 6.5).

Figure 6.5. Plantations to extract ethanol.

Source: Image by Wikipedia.

Although, biodiesel and ethanol when burned cause less amount of carbon than petroleum but on the other, the process of formation of these biofuels leads to these changes in the environment. In order to find new energy sources and ways to convert biomass into usable energy and as the result of the above-mentioned effects of biofuel formation environment has encouraged scientists, politicians, and the public are looking to genomics.

Further, the second generation biofuel production basically focuses on the extraction of energy from lignocelluloses biomass sources, and these advances in these areas are observed due to the efforts in this field (Figure 6.6).

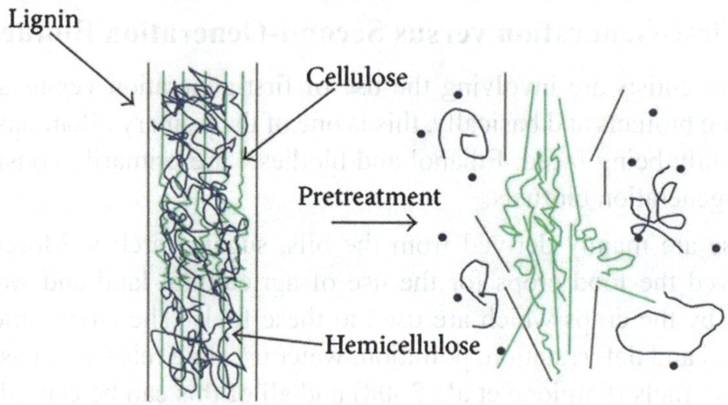

Figure 6.6. Deconstruction of lignocelluloses into cellulose.

Source: Image by Researchgate.

Lignocelluloses Biomass consists of matter composed of the woody, inedible parts of plants, such as grasses, crops, and forest waste unlike the easily processed sugars and oils in the first-generation biofuels.

The removing competition between the use of this land for food production and the use of this land for fuel production and this is one of the greatest advantages of these materials. And this competition is only removed because these materials are not typically grown on the agricultural land.

Basically, the major question which arises due to the above context is that how is energy stored and extracted from lignocelluloses biomass? Firstly, the solar energy is converted by the plants by the process of photosynthesis. This converted solar energy is stored in the cell walls polymers. This energy is stored in the forms of cellulose, hemicelluloses, and lignin (Rubin, 2008).

Unlike starch, cellulose is a chain of glucose molecules and it is not soluble in water. This cellulose is basically the main fuel component in lignocelluloses biomass.

Further lignocelluloses biomass which is basically celluloses comes bonded to the lignin (this provides the plant structure and natural defense against the insects and the pathogens) and the other is hemicelluloses (this basically consists of five and six carbon polysaccharides) and this bond is basically another challenge in using the lignocelluloses biomass.

Various enzymes present in yeasts and fungi, various celluloses and hemicelluloses have been easily identifying and moreover are successfully enhanced the degradation of cellulose (Turner et al., 2007) although hemicelluloses are more difficult to ferment than cellulose.

Lignin itself immune and protects both the cellulose and hemicelluloses from enzymatic conversion, and however, this is the primary obstacle to saccharification.

First-generation feedstocks like corn, sugarcane, soy, palm, or grape seed can be much easily degraded as compared to the lignocelluloses biomass as inferred by the above context. So, further, to improve the yields of this plentiful and potentially inexpensive fuel source in turn researchers are continuously seeking methods for refining the process.

6.4.2. Finding the Ideal Source of Biomass

For biofuel production center on fuel stocks with modifiable traits, the main areas of genomic investigation in optimizing lignocelluloses biomass. This biofuel production can be engineered for biorefining (Sarath et al., 2007). Some of the best options for bioethanol feedstocks are C4 perennial grasses such as switchgrass, sorghum, and Miscanthus.

Due to the yield rate and the ability to grow from seed on marginal lands, the capacity to protect soil from erosion and potential to grow without fertilizers or replanting (Bouton, 2007) the U.S. Department of Energy's (DOEs) BioEnergy Feed Stock Development identified switchgrass as one of the most promising energy. And this is one of the examples.

The researchers have already observed that although many genotypes of switchgrass exist, only two cultivars present options for efficient genetic engineering (Sticklen, 2008), but The Joint Genome Institute is currently sequencing the switchgrass genome.

Trait analyses must be conducted for other biomass sources which include plants such as C3 plants like poplar and willow as it is observed that the switchgrass cannot be grown in the regions of cold climates.

Researchers are still at work on maximizing poplar's speed of growth, cell wall chemistry, branching tendencies, the thickness of stems, and reaction to competition for light (Rubin, 2008) although poplar was considered as one of the first bioenergy sources to be sequenced.

A model for C4 fuel stocks and sequencing of the B73 corn cultivator could nearly complete by the corn, or maize and this many researches also think. A final genome to be completed by the end of 2008 and this was announced by Washington University in St. Louis in February 2008 that they had released a draft blueprint.

A long source for key scientific breakthroughs such as detection and cloning of transposable elements, the correlation of cytological and genetic crossing over, and also the uncovering of several other genetic phenomena (Lawrence et al, 2008) that not only maize is the foremost ethanol fuel stock in the U.S.

As maize is similar to switchgrass and miscanthus in using the C4 photosynthetic pathway so maize is basically a particular attractive model for genetic modification of perennial grasses. Maize also originates in the same subfamily of the PACCAD clad (Panicoid, Arundinoid, Chloridoid, Centothecoid, Aristidoid, and Danthonioid lineages).

In a similar manner to the miscanthus maize can be easily be hybridized moreover this is due to the corn subspecies which includes diploids and tetra plods as they are both annual and perennial (Lawrence and Walbot, 2007).

6.4.3. Third-Generation Biofuels from Microalgae and Cyanobacteria

The so-called third-generation biofuels were also being worked upon by the scientists beyond the second generation. These third-generation biofuels are basically derived from microalgae and Cyanobacteria. Further, as extensively as lingo cellulose biomass, this field is not yet been explored.

The sunlight, water, and CO_2 to their oil by the photosynthetic organism's microalgae and later on, through existing refining processes which when they are converted to biodiesel. The fact that the microalgae are an aquaculture which reproduces rapidly and this is one of the tremendous advantages to this process including all this. Microalgae also has various other characteristics as they result in biofuel can be used as a pure product in existing transport systems, they can also grow in saltwater, and they also generate a biomass with a higher energy content as compared to the ethanol (Holzman, 2008) (Figure 6.7).

Figure 6.7. Cyanobacteria can be a good source for biofuel.

Source: Image by Flickr.

Critical factors such as sunlight, CO_2 levels are considered for the actual amount of the oil produced. Development of costly and researcher related to land use options for pond farms and as yet imperfect photobioreactors are the results which are required for the research.

Increased oil content in the biomass, and regulated temperature tolerance, which is a current challenge moreover, an increased photosynthetic and biomass yield, advanced growth rates are the result of the genetic engineering of microalgae (Chisti, 2007).

Microalgae and Cyanobacteria which are also known as the blue-green algae basically help to develop biohydrogen which is another type of fuel. For completely environmentally clean fuel this fuel is popular and one of the tremendous advantages of this fuel.

Fossil fuels and release green gases into the atmosphere are still today used to create the biohydrogen. The formation simply relies on the GHGs release and fossil fuels. Further, it was discovered that unicellular microalgae could produce hydrogen, and this was basically discovered by the researcher named Hans Gaffron at the University of Chicago.

After that, the researcher named Kruse et al. (2005) discovered that hydrogenases across the prokaryotic and eukaryotic realms. The solar energy was used to split water (bio photolysis) and generation of hydrogen without releasing any CO_2 alike the microalgae and cyanobacteria hydrogenases which use them.

The presence of oxygen tends to generate lower yields of hydrogen is however acting as a major drawback under the examination. So, basically to overcome this drawback genetically engineering will be necessary to maximize hydrogen production. To oxygen, the hydrogenase's tolerance was adjusted.

The development of sustainable bioenergy and energy conversion processes these are just a few ways that genomics is influencing. Moreover, there is a need to explore many other genetic solutions. Further to deal with the global crisis in the upcoming years the researchers definitely will continue to look towards nature for more solutions (Figure 6.8).

Figure 6.8. Microalgae can be extremely efficient sources of energy.

Source: Image by Pxfuel.

Scientists can optimize billions of years of evolution to meet our growing needs in an environmentally friendly way, this was done by applying genomic research and engineering to renewable fuel stocks and the bacteria and the enzymes which convert those sources to energy.

6.5. THE CHEMISTRY OF BIO FUELS

The chemistry of biofuel is more complex as compared to the petroleum and this is mainly stated according to the researchers at Lawrence Livermore National Laboratory and Sandia National Laboratory. For this enhanced complexity of the above statement there are mainly two reasons and these reasons are basically explored here.

The properties of biofuels are looked closer by the means of the subsequent articles. A greater expertise in chemistry and biology is required for producing the biofuels and this requirement is the bottom line of this entire concept. Further, a greater knowledge in biology and chemistry is not required when there is dealing with the fossil fuels.

6.5.1 Variety

Due to the presence of almost all the molecules of hydrocarbons petroleum is rather considered as a simple substance. It is observed that they occur in an altering length. Further, they can be in any of the state some of them can be liquid state, others can be in solid and some can be gases as well.

All of them basically resemble means they almost behave similar to each other due to the fact that all of them are made up of hydrogen and carbon. For biofuel the same is not true.

Oxygen is basically present in the biofuels. Dramatically, it has been observed that the complexity of these molecules started rising due to that one atom. The presence of the single or double present is the basic measure which further helps to understand the major difference between the fossil fuels.The presence of this major classification around which this difference fossil fuels revolves around. While, on the other hand in case of biofuels it is more difficult to figure out the differences. Biofuels can also contain esters, ethers, alcohols, and acid groups and this is possible only due to the presence of oxygen in the biofuel.

A whole subset of organic chemistry is created by each of these groups. They possess special reaction characteristics. Esters are not like acids at all really and moreover, alcohols are also not like alcohol. A huge jump in complexity is observed as the net result of adding oxygen.

6.5.2. Biology and Synthesis

The processes are all chemical or physical, when dealing with fossil fuels. As fossil fuels are more refined so, basically the fossil fuels are not really produced so much. Many useful compounds are already present in the petroleum such as gasoline, diesel, and others.

In order to separate out the parts it is observed that all industries basically refine the petroleum. The basic process of getting fuel from petroleum is relatively straightforward but other processes are also involved in this like cracking and these are basically done to improve the yield.

Biofuels do not goes under the same processes. Firstly, the feedstock has to be produced, before a biofuel can ever be used. A complex understanding of biology and chemistry is required in such processes. To reduce the nutrients and water requirement and also to improve the yield in many organisms are genetically modified.

Even before planting a biofuel crop therefore, one must undergo through a grasp of biology. Algae are used for production after that biofuel is either refined from oil or produced and harvested. To obtain useful fuels out of biological molecules there is a need of complex chemistry involved in either of the cases.

6.5.3. Diesel and Biodiesel

From the fractional distillation of crude oil, a product is produced which is termed as petroleum diesel or petrodiesel. A mixture of hydrocarbon molecules is contained in the product and these molecules are in the ranges between the sizes from 8 to 12 carbon atoms. 16 carbon molecules like a structure are basically observed from the typical petrodiesel molecule.

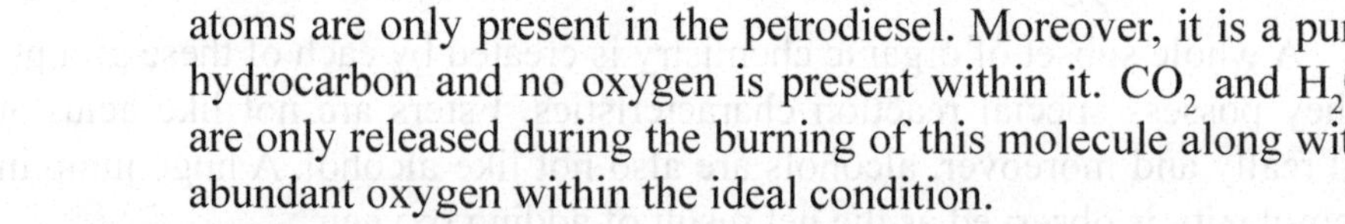

1. **Petro-Diesel with 16 Carbon Molecules:** Hydrogen and carbon atoms are only present in the petrodiesel. Moreover, it is a pure hydrocarbon and no oxygen is present within it. CO_2 and H_2O are only released during the burning of this molecule along with abundant oxygen within the ideal condition.

Small amounts of gases like CH_4, neon, helium, hydrogen, etc., are present in the atmosphere and moreover, the major portion in the atmosphere is covered by 21% of oxygen and 78% of nitrogen. Hydrogen sulfide (H_2S) and other sulfur compounds are used to contaminate many petroleum deposits beyond that.

Sulfur compounds like sulfuric acid and nitrogen compounds such as nitric acid and other contaminants are basically the results which are produced by the burning petroleum. The reaction which is produced cannot be considered as ideal on the other hand.

With the help of a typical biodiesel molecule, the petrodiesel molecule is being compared by the below-mentioned structure.

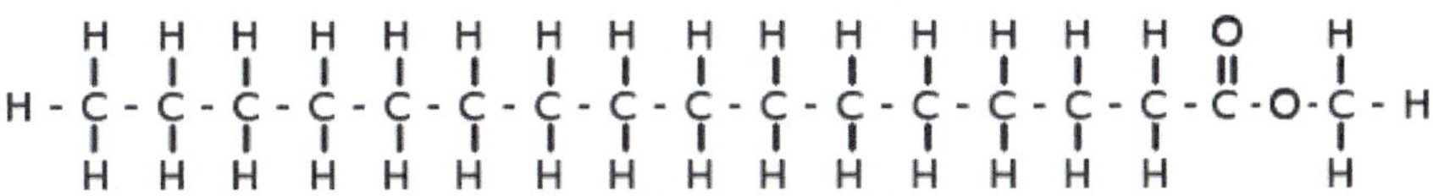

2. **Biodiesel with 17 carbons (also called 16 carbons with an ester group):** Further it is observed that the petrodiesel molecules and the biodiesel are similar to each other in many ways.

There are many similarities between the biodiesel and petrodiesel molecules. As compared to the petrodiesel molecule the only difference is observed in the biodiesel is on the right side of the molecule which has two oxygen atoms. When these are burned, the difference in the biofuels like biodiesel is only due to these oxygen atoms.

6.5.4. Why the Oxygen Exists in Bio Fuels?

The way the oxygen is produced is the reason of the presence of the oxygen in biodiesel. Over the long periods of time and under anaerobic conditions the petrodiesel is produced. From the dead plants and animal matters, these conditions result in the removal of the oxygen. To form petroleum and other fossil fuels the conditions only leave hydrogen and carbon.

Through the process known as transesterification on the other hand helps in creation of biofuels. To yield very long molecules fats and oils from living organisms are broken apart in this process. These long molecules basically contain oxygen that is seen below.

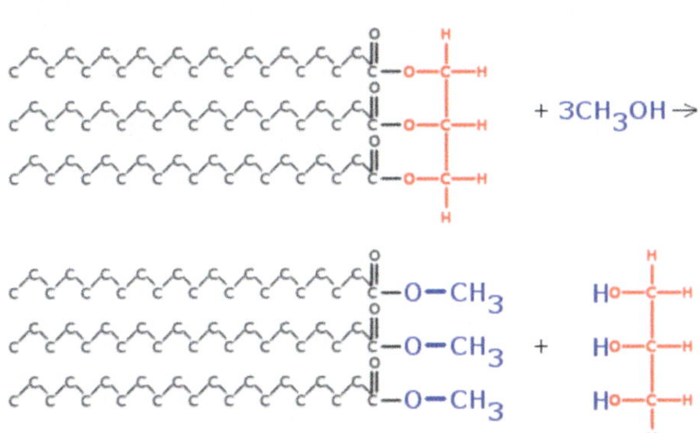

1. **Transesterification of Fatty Acid (glycerol shown in red, methanol shown in blue):** With slightly different qualities than petrodiesel the molecules which are produced of biodiesel is produced in this way. A methanol molecule is taken (shown in blue color) and in essence, a triglyceride molecule is taken (made up of three fatty acids and a glycerol molecule that are shown in red) these two are reacted together.

This process basically leads to a production of a mixture which basically contains fatty acids and esters (blue and black). Further, this created mixture can be used in chemical applications, food processing, and other things and this mixture also be used glycerin (blue and red) and as well as biodiesel.

6.5.5. Ethanol, Butanol, and Gasoline

Organisms can produce ethanol and butanol on their own and it is the biggest difference between these biofuels (ethanol and butanol) and biodiesel. To obtain a raw product which is converted into biofuels, there is no harvest an organism. Fewer nutrients and less water is required is the major advantage of such an approach.

The diagram which is drawn below is the molecular structure of ethanol, butanol, and gasoline. The number of carbon atoms contained in the ethanol and butanol is the major difference between these two and this is basically a structural difference. Although gasoline is a mixture of hydrocarbon molecules which basically consists the ranging in size from four to twelve carbon atoms long but still it is noted that octane is used to represent gasoline.

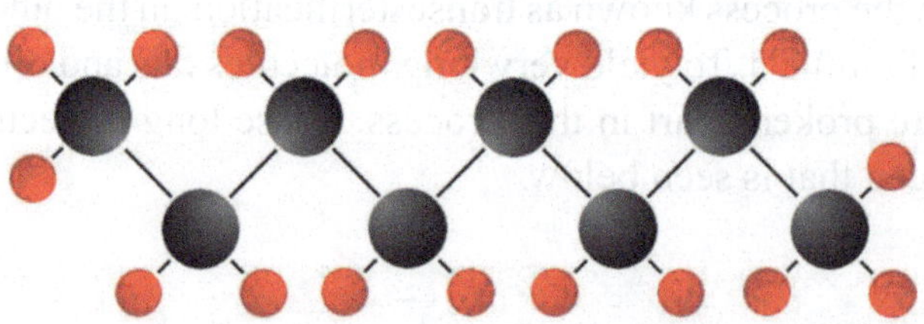

1. Octane, one of Several Molecules in Gasoline

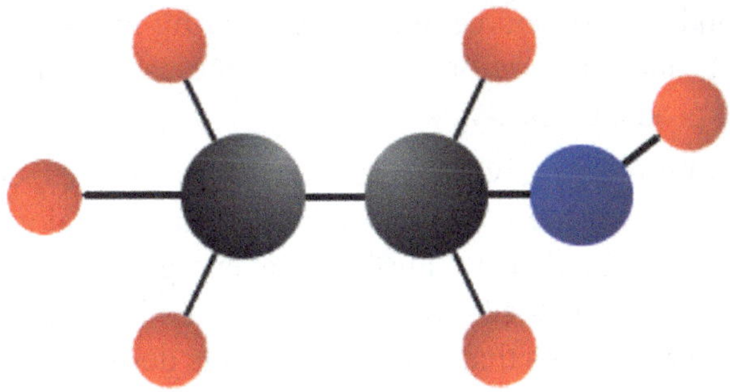

2. Ethanol (blue represents oxygen)

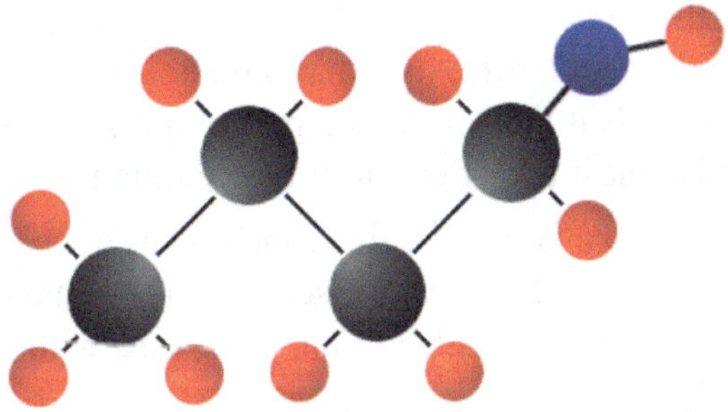

3. Butanol (blue represents oxygen)

As it is observed that ethanol and butanol can be produced from the living or recently living organism, therefore, ethanol, and butanol are both considered as biofuels. In the presence of oxygen there is a major difference between ethanol and butanol and gasoline similarly as there is a difference between petrodiesel and biodiesel.

There is alcohol group (just an oxygen with hydrogen attached) but there is no ester group in case of alcohols (ethanol and butanol are alcohols). Esters that make up biodiesel basically behave differently as compared to the alcohols.

6.5.6. Ethanol Versus Butanol

Further, it is clear that gasoline and butanol are similar to each other than in ethanol. Butanol is not been reliably demonstrated on a large scale that and also producing butanol is more difficult. As a part of their normal chemistry ethanol is being produced by many organisms, these organisms are namely yeast and even by some types of algae.

Just like oxygen is produced with the help of plants the above context infers that similarly ethanol is produced by number of organisms. To produce ethanol CO_2 and energy source are just needed by many organisms. Only water, sunlight, and CO_2 are only needed in the case of algae. Therefore, ethanol can be produced easily.

Ethanol is highly corrosive to the engine component, especially rubbers, and it takes a deal of energy to produce and this is major problem with ethanol, beyond energy density. Further, this energy could be offset by the means of changing feedstock from corn to something like Jatropha or by using algae. Later, it seems that besides ethanol is derived from it but so far, more amount of energy is required. Biodiesel is produced by the means of algae.

As butanol takes the same amount of energy to produce similar to ethanol due to this butanol is less corrosive than ethanol. Further in long run, it seems more efficient. But there is also a problem regarding the production of butanol, and it is that it is difficult to produce butanol at a large scale.

To produce butanol in large quantities and on large scale recently several companies have devised plans. And also, to make butanol commercially viable fuel these companies are also moving forward with further plans.

Both ethanol and butanol are biodegradable. They are more stable as compared to the biodiesel during long term storage; also, unlike its ethanol and butanol do not freeze at lower temperatures. Therefore, the above context infers that better offers are provided by these fuels as an alternative to gasoline, within in the long run, and as compared to petrodiesel, biodiesel offers.

6.6. THE FUTURE

In the production of biofuels, there is a need of understanding biology in order to assist this production. Things like algae used to produce fuel this have been induced by some companies and this is done by the means of genetically modifying organisms.

The fuel produce is so similar to each other that it might be difficult to figure out the difference. The production of fuel can actually reduce carbon emissions if it could be possible to complete this process without investing a great deal of energy from standard fuels (Figure 6.9).

Figure 6.9. Biofuel research laboratory at Sandia.

Source: Image by Sandia Energy.

In this way, the energy for the production of fuel could be obtained by the sunlight. To produce the fuel the carbon is taken up and this is the reason the process occurs. This produced fuel is then released when it is burned. It is theoretically possible but still, there is a holy grail of biofuel production that has yet to be obtained.

In the current scenario, 9.5% of total primary energy supply and some 70% of renewable energy the bioenergy is a significant part of the energy economy (International Energy Agency, 2017b, 2019). Also, within the small industries (such as charcoal kilns and brick kilns) more of the traditional use of biomass, mostly in households for cooking and heating are used.

This traditional biomass use is basically taking place in more than half of the bioenergy. With the use of such traditional biomass, there is also considerable scope for improving various aspects. These aspects include the sustainability, health safety, and efficiency (Creutzig et al., 2015). Because of its potential for significant growth in the coming decades this paper limits its focus to modern bioenergy.

Four times the contribution of solar photovoltaic (PV) and wind combined (International Energy Agency 2018) was basically contributed by the modern bioenergy (hereafter bioenergy) which was responsible for half of all renewable energy consumed in 2017.

Further, the bioenergy is also expected to account for 3% of electricity production and around 4% of transport energy demand in 2023 (International Energy Agency, 2018) also most of the bioenergy delivers heat in buildings and industry.At annual rates greater than 10% prior to 2010, the production of liquid biofuels for transportation grew. Later, from 2010 to 2016 the annual growth of production of liquid biofuels slowed down to 4%.

From 2010 to 2016 (International Energy Agency, 2017b) the annual average growth rate of bioenergy electricity capacity was 6.5%. The growth in renewable energy production (International Energy Agency, 2018) the bioenergy (including liquid biofuels) projected to account it for 30% over the period of 2018 – 2023.

For addressing climate change in the recent scenarios, the bioenergy features predominantly used. The biomass was made up of a median of 26% of primary energy (154/EJ /year) in 2050 (range from10% to 54%) up from 10% in 2020 (Rogelj et al., 2018) and this was basically founded by the recent Intergovernmental Panel on Climate Change (IPCC) Special Report on Global Warming of 1.5 degree Celsius and it also reviewed eighty five1.5 degree Celsius pathways.

With a significant growth in bioenergy with CCS (BECCS; Rogelj et al., 2018) and use of high levels of bioenergy without carbon capture and storage (CCS), there assessment was basically reviewed by the means of many modeled pathways.

With a reduced amount of bioenergy, some of the modeled pathways achieve a 1.5 degree Celsius without BECCS. Significant changes in behavior (such as changes in diet), rapid technological progress, and a low global production is the means of which it based upon also on a substantial decline in overall energy demand (a reduction of 32% by 2050 compared to 2010 levels).

A keen interest worldwide is obtained by the near records of the oil price. The plant material is used to make the biofuel and it is observed as the future of the biofuel. The global biofuel production is less than 3% of the global transportation fuel supply but significantly the production of global biofuel has been tripled from 4.8 billion gallons in 2000 to about 16.0 billion in 2007.United States, Brazil, and the EU about 90% of production are concentrated in these areas. In other countries such as Malaysia and China if the development programs are successful it could in turn lead to the production to become more dispersed. Corn, sugar, and vegetable oil are the leading raw materials or feedstocks for producing biofuels.

There is an expectation about the potential substitutes for the oil-based fuels, apart from the rapid expansion in biofuel production. Further the commodity prices on the global food system were a matter of concerns as there could be a price rise due to this.

Due to the increase in corn, wheat, and soybean prices it was observed that the food price rose 10% in 2006 and this was stated by according to the International Monetary Fund. The biofuel demand was also included primarily from the demand side factors.

Due to rising feed prices, a moratorium on expanded use of corn for ethanol was made by the Chinese government. The food crops such as cassava, sweet sorghum, and jatropha (an oil-bearing plant originally from South America) do not compete directly by the other feedstocks.

To contain food price inflation from higher-priced corn imports in early 2007 Mexico capped tortilla prices. In Brazil and elsewhere there was a stressing budget of low income people in 2006 as the real sugar prices hit a 10 year high. Since, then a decline in prices was observed. Further, to slow down the rising cost of domestic cooking oil in mid-2007 the export duty on crude palm oil was increased by the Indonesian government.

Higher retail meat prices might be translated by the increased costs for corns and other feed faced by the US livestock producers. The complete dependence on imports of fed grain and oilseeds to support its large livestock sector and this was historical concerns have been revived and this country was Japan.

The future price of oil, availability of low-cost feedstocks, sustained commitment for supportive policies by governments, technological breakthroughs and these are factors for the outlook for the global biofuels and it is that these factors inter-related. The competition from unconventional fossil fuel alternatives and could reduce the cost of second-generation biofuels.

6.6.1. Looking to the Future: The Potential of Second-Generation Bio Fuels

The competition from fossil fuel alternatives and concerns about the environmental tradeoffs were included by many uncertainties remain for the future of biofuels. The current biofuel production can be reduced by the means of land intensity, which is perhaps the biggest uncertainty. An acre of land varies from 100 gallons per acre for EU rapeseed to 400 gallons per acre for US corn and 660 gallons per acre for Brazilian sugarcane will produced the amount of biofuel.

The land requirements were significantly reduced by the cellulosic ethanol which will rise per acre ethanol which yields to more than 1000 gallons. The tough cellular material gives rigidity and structure, and this is broken, and cellulosic ethanol is made by the breakage. This cellulosic ethanol also helps to convert the resulting sugar into ethanol.

Wood chips, wood waste, fast-growing, crop residues like corn Stover, and municipal waste cellulose is present in the low-value materials and moreover, cellulose is the most widely available biological material.

Compared with 1.65 dollars per gallon for corn ethanol and 2.50 per gallon this is basically estimated costs now and US cellulosic fuel production. Several other countries such as Canada, Brazil, China, Japan, and Spain apart from the US the venture capital and government subsidies were supported companies interested in making cellulosic ethanol.

Impacts of harvesting grasses, trees, and crop residues on the erodibility and fertility of land resources these were the other costs of cellulosic ethanol production fully assessed within the meantime. In the processing of the storing large volumes of bulky feedstocks, transporting, and the environmental costs of harvesting these are major questions regarding the upstream logistical.

6.6.2. Future Role of Bio Fuels Depends on Profitability and New Technologies

The economic cost and the environmental impacts for biofuel production could be steadily reduced by the means of the higher biomass yield per acre and more gallons of biofuel. And these biomasses were mainly due to the efficiency gains and the technological advances.

In tropical areas where growing seasons are longer, per acre biofuels yields higher the biofuel production will likely prove to be most profitable and environmentally benign. These tropical areas also posse's fuel and other input cost lower.For example, bagasse is used by Brazil. To power ethanol distilleries the bagasse is a byproduct of sugar production. Natural gas or coal is used in a country like the US. The profitability which depends on number of interrelated factors the future of biofuels depends on them.

The economic support for alternative fuels, unlike previous periods the key factors which occurred were the high oil prices: 6 years of steadily rising oil price. There was an undercutting the profitability of nascent alternative fuel programs due to the oil prices which spiked in the previous periods.

6.7. CONCLUSION

Bioenergy is an important source of energy in the world that looks forward to a sustainable future. Biofuels play an important role in this regard as they are the main source of bioenergy. The biofuels are the fuels that are manufactured from burning of plants and trees.

The biofuels are currently being used in the form of biodiesel and ethanol. The conversion process of the biofuels is a simple one with some important steps. They are important to remember, in the context of bioenergy generation.

Biofuels are contributing significantly to developing and maintaining the sustainability in the energy industry and has a major role to play in the context of renewable sources of energy. The future of biodiesel is extremely bright and holds a lot of relevance and importance as the world looks forward to such sources of energy.

REFERENCES

1. Biofuel.org.uk, (n.d.) Biofuels-Chemistry of Biofuels-How Are Biofuels Produced? [Online] Available at: http://biofuel.org.uk/how-are-biofuels-produced.html (accessed on 4 July 2020).
2. Biofuel.org.uk, (n.d.) Biofuels-Chemistry of Biofuels. [Online] Available at: http://biofuel.org.uk/chemistry-of-biofuels.html (accessed on 4 July 2020).
3. Cockerill, S., & Martin, C., (2008). Are biofuels sustainable? The EU perspective. Biotechnology for Biofuels, [Online] 1(1), 9. Available at: https://biotechnologyforbiofuels.biomedcentral.com/articles/10.1186/1754-6834-1-9 (accessed on 4 July 2020).
4. Coyle, W., (2007). USDA ERS-The Future of Biofuels: A Global Perspective. [Online] Ers.usda.gov. Available at: https://www.ers.usda.gov/amber-waves/2007/november/the-future-of-biofuels-a-global-perspective/ (accessed on 4 July 2020).
5. Davidson, S., (2008). Sustainable Bioenergy: Genomics and Biofuels Development | Learn Science at Scitable. [Online] Nature.com. Available at: https://www.nature.com/scitable/topicpage/sustainable-bioenergy-genomics-and-biofuels-development-44571/ (accessed on 4 July 2020).
6. Eesi.org, (n.d.) Bioenergy (Biofuels and Biomass) | EESI. [Online] Available at: https://www.eesi.org/topics/bioenergy-biofuels-biomass/description (accessed on 4 July 2020).
7. Energy.gov, (n.d.) Biofuels Basics. [Online] Available at: https://www.energy.gov/eere/bioenergy/biofuels-basics (accessed on 4 July 2020).
8. EU Science Hub-European Commission, (n.d.) Biofuels and Bioenergy-EU Science Hub-European Commission. [Online] Available at: https://ec.europa.eu/jrc/en/research-topic/biofuels-and-bioenergy (accessed on 4 July 2020).
9. Reid, W., Ali, M., & Field, C., (2019). The future of bioenergy. Global Change Biology, [Online] 26(1), 274–286. Available at: https://onlinelibrary.wiley.com/doi/full/10.1111/gcb.14883 (accessed on 4 July 2020).
10. The Guardian, (n.d.) What are Bioenergy and Biofuels–And are They a Good Idea? [Online] Available at: https://www.theguardian.com/environment/2011/sep/23/bioenergy-biofuels-climate-change-faq (accessed on 4 July 2020).

Chapter 7 Green Chemistry and Sustainable Energy

CONTENTS

Green chemistry, also called sustainable chemistry, is an area of chemistry and chemical engineering focused on the designing of products and processes that minimize or eliminate the use and generation of hazardous substances. This chapter discusses different paradigms of the green chemistry and its advantages and disadvantages are also discussed. Initially, the chemical industry and the process of development are discussed considering sustainable approach. After that, the principle and application of the green chemistry is evaluated. At the end, the skills of green chemistry is discussed, and the biocatalyst are also described.

7.1. INTRODUCTION

A new paradigm that promises to have a deep and lasting impact on the science of chemistry is known as "green chemistry" and "sustainability" and now there is not just considered as the sound bites. It is said that green chemistry and sustainability essentially go hand in hand.

Meeting the needs of the present generation without compromising the ability of future generations to meet their own needs is known as sustainable development. There is the requirement of greener chemistry.

Because with the help of greener chemistry, there will be efficient utilization of renewable raw materials, and with the help of this use of toxic can be avoided and waste can be eliminated and processes in order to achieve this noble goal and or hazardous solvents and reagents in both products.

There are two main components which are incorporated by green chemistry. Firstly, the problem of related elimination of waste and the problem of efficient utilization of raw materials is addressed.

Secondly, the main issue is the environmental issue which is associated with the manufacture, use, and disposal or re-use of chemicals and it also deals with the health and safety. To use on the path to sustainability the green chemistry is considered as one of the most fundamental and powerful tools. In fact, it is said that without green chemistry and green engineering, there is no path to sustainability.

There are some manufactures, who are manufacturing a large number of so-called commodity chemicals, such as nylon, adipic acid, phenol, and a variety of epoxides and esters, and these manufactures producing these chemicals on the scale of many millions of tons per annum and also there are several environmentally unacceptable features.

While in the process of manufacturing there are most of these materials consume excessive energy. The waste which currently disposed of landfills is generated as they also entail the use of aggressive, corrosive, and sometimes explosive reagents.

It has been already known that too many individual steps are usually required in the production of fine chemicals and pharmaceuticals and a majority of these are stoichiometric, thereby consuming large amounts of reagents.

So, there is a need to make the plan regarding the new methods of manufacturing which will help in to minimize the consumption of energy and materials (including volatile solvents) and that also minimize the liberation of harmful gases (typified by the notorious GHG, N_2O) is pressing.

Therefore, after the identification of the requirement of the new plans progress has been made to utilize specially designed catalysts that comply with the above requirements and also function under much more benign reaction conditions [e.g., with the use of O_2 (or air) or H_2O_2 (or an alkyl peroxide)] in the place of aggressive oxidants such as HNO_3, $K_2Cr_2O_7$, or $KMnO_4$.So, after this replacement of steps is required, each with a low product yield, in the manufacture of pharmaceuticals by a few catalytic processes is urgently called for.

7.2. SUSTAINABLE DEVELOPMENT OF THE CHEMICAL INDUSTRY

There are many manufactures of the products which involve one or more chemical processes. So, it is impossible to imagine that how the life will be like without the products produced by the chemical industry.

However, it has been seen that the resources taken from the earth, on average it is in a small proportion of the resources and it is converted into the desired products in current chemical processes, and after that large amounts of wastes and hazardous materials are generated.

So, there is always question arise which is an unavoidable issue, that how to supply humanity with enough food, energy, chemicals, and materials sustainably, without damaging our planet.

Two decades ago, green chemistry was established, which has attracted much attention. Challenges which are related to green chemistry are reflected by the efforts of academia and industry, and continuous progress is being made, both in academia and industry (Figure 7.1).

Figure 7.1. Sustainable chemistry is required for industries.

Source: Image by Wikipedia.

Briefly, to reduce or eliminate the use or generation of hazardous substances in the design, manufacture, and applications of chemical products, green chemistry is the utilization of a set of principles. Multidisciplinary field is considered as the green chemistry and covers areas such as synthesis, solvents, catalysis, raw materials, products, and efficient processes.

7.2.1. Efficient Synthetic Routes

Large amount of waste is produced because most of the current chemical production processes lack efficiency in using feedstock's. For reducing both the depletion of raw materials and the generation of waste, it is required to increase atom economy and it is very crucial.

Ideally, it has said that all the atoms in reactants should be transformed into the desired products. However, it has been observed that it is not realistic to achieve 100% atom economy in all industrial chemical processes.

Integration of different reactions and processes is another way to eliminate waste, in which the by-product in one reaction is the feedstock of another. To increase the synthetic efficiency and reduce or eliminate wastes is a long-term task, as because of the exploration of atom-economic synthetic protocols and routes.

7.2.2. Greener and Functional Solvents

To prepare chemicals and materials there is the requirement of huge amounts of toxic, flammable, and volatile organic solvents are used in chemical processes.

It has been observed that each year about 20 million tons of organic solvents are released to the atmosphere, which is leading to solvent waste and environmental pollution.

There some solvents like water, supercritical fluids, ionic liquids, non-toxic liquid polymers which are known as the greener solvents which are being used and their various combinations in chemical processes has become a major focus of research in academia and industry.

It is required that a green solvent should meet some basic requirements such as low toxicity, ease of availability and recycling, and high process efficiency. It is said that properties of the solvents which are used, as the efficiency of a process usually depends strongly on this.

As the special properties are there, so because of its properties and functions to optimize chemical processes green solvents can be used, decrease solvent usage and processing steps, and develop new routes and technologies that meet the requirements of sustainability.

7.2.3. Greener Catalysis

A key role is played by the catalysis in the chemical industry because to accelerate reactions most chemical processes need catalyst; enhance selectivity and lower energy requirements. Expensive, toxic, harmful, or noble metals are the one current catalyst is often based on.

High activity, selectivity, and stability, and ease of separation and reuse are some common characteristics which green catalysts should have, and it is said that environmentally benign and widely available raw materials such as abundant metals, organic compounds, and enzymes are the one they should be based on.

The exploration and development of new synthetic routes and chemical processes rely strongly on progress in catalysis. To achieve the dual goals of environmental protection the design is used of green catalysts and catalytic systems and economic benefits are an important task and are essential for the sustainability of the chemical industry.

7.2.4. Use of Green and Renewable Feedstock's

Fossil resources are the one on which current energy supply and the feedstock are mainly based for producing organic chemicals and materials, which are said to be not renewable and are diminish. So, there carbon resources like biomass and CO_2 and the use of it in the chemical and energy industries is extremely important, and different routes and processes have been developed.

Nonetheless, we face thermodynamic, kinetic, and specialized difficulties in the change of biomass and CO_2 into fills and synthetic concoctions. Numerous present courses are actually plausible, yet monetarily restrictive, and without a doubt, extremely little extents of the assets are as of now utilized.

The improvement of effective techniques for changing over biomass and CO_2 into helpful synthetics and liquid powers through vivaciously and financially feasible mechanical procedures is vital yet is testing.

In addition, the utilization of greener, less expensive, more secure reactants and sustainable energy sources, for example, oxygen, hydrogen peroxide, and sun-based energy, in compound procedures is likewise an intriguing territory.

7.2.5. Green Engineering and Products

Engineering aspects and the green products are the one on which is covered by the green chemistry. Clean, energy-efficient, and mass-efficient processes and technologies are essential tools for achieving the goal of maximizing efficiency and minimizing wastes, as chemicals and materials are produced by industrial chemical processes.

There are many chemicals which are harmful like pharmaceuticals, fine chemicals, commodity chemicals, and polymers. To replace hazardous products, products that are benign to human health and the environment need be designed and produced.

It has been clearly observed that the exploration of synthetic routes, design of sustainable products and solvents, and exploration of new catalysts and chemical processes are very much closely related, and they should be integrated. In addition, central driver for the development of green chemistry and technology are the economic benefits (Figure 7.2).

Figure 7.2. Green engineering for product development.

Source: Image by Flickr.

In future green chemistry will be one of the most important fields. Even though, it has been that in the last 20 years this field has developed rapidly, and it is still at an early stage. As everyone know that promoting green chemistry is considered as the long-term task, and along with this there are many challenging scientific and technological issues which are need to be resolved and these are related to chemistry, material science, engineering, environmental science, physics, and biology.

To promote the development of these field scientists, engineers, and industrialists should work together. As there is no doubt in this that development and implementation of green chemistry will contribute greatly to the sustainable development of our society.

7.3. GREEN CHEMISTRY AND SUSTAINABILITY

It has been observed that much attention is attracted by the green chemistry, which was established about two decades ago. To address the challenges related to sustainable development of the chemical industry it reflects the efforts of academia and industry, and there is continuous improvement, which is being made, both in academia and industry.

Briefly, it has been described that to reduce or eliminate the use or generation of hazardous substances in the design, manufacture, and applications of chemical products, green chemistry is the utilization of a set of principles. Areas are covered like synthesis, solvents, catalysis, raw materials, products, and efficient processes; green chemistry is referred as the multidisciplinary field.

Phasing out the use of chlorofluorocarbons (CFCs) in refrigerants are some of the important examples of green chemistry. As it has played very important role in creating the ozone hole; developing more efficient ways of making pharmaceuticals, including the well-known painkiller ibuprofen and chemotherapy drug Taxol; and developing cheaper, more efficient solar cells.

The basis of vast multinational industries is making chemical compounds, particularly organic molecules (composed predominantly of carbon and hydrogen atoms), and it is from perfumes to plastics, farming to fabric, and dyes to drugs.

In one practical these would be prepared from inexpensive, renewable sources, efficient, safe, and environmentally benign chemical reaction. Unfortunately, the majority of chemical processes are not completely efficient, as with the exception of the chemical processes found in nature and it require multiple reaction steps and generate hazardous by-products.

As it has been observed earlier also that in the past traditional waste management strategies are completely focusing only on the disposal of toxic by-products and today the efforts have been made which are shifted to eliminating waste from the outset by making chemical reactions more efficient.More sophisticated and effective catalytic reactions are led by the advent of this adjustment, which reduce the amount of waste. The 2001 Chemistry Nobel Laureate Ryoji Noyori stressed that catalytic processes represent "the only methods that offer the rational means of producing useful compounds in an economical, energy-saving, and environmentally benign way."

In the future green chemistry will considered as the one of the most important fields. It has been observed that this field has been developed rapidly in the last 20 years, and it is still at an early stage.

Long term task is considered for promoting green chemistry, and there is requirement to resolve the issues of many challenging scientific and technological issues, and these are related to chemistry, material science, engineering, environmental science, physics, and biology.

The development of this field is only possible when the scientists, engineers, and industrialists will work together to promote the development of this field. It has been said that there is no doubt that the development and implementation of green chemistry will contribute greatly to the sustainable development of our society.

7.3.1. Principles of Green Chemistry

- Prevention: After the waste has been created, it is better to prevent waste than to treat or clean up.
- Atom Economy: Materials used in the process into the final product is the incorporation of all materials, as synthetic methods should be designed to maximize these.
- Less Hazardous: To use and generate substances that possess little or no toxicity to human health and the environment chemical syntheses wherever practicable, synthetic methods should be designed.
- Designing Safer Chemicals: While minimizing their toxicity chemical products should be designed to affect their desired function.
- Safer Solvents and Auxiliaries: The use of auxiliary substances (e.g., solvents, separation agents, etc.) should be made unnecessary wherever possible and innocuous when used.
- Design for Energy Efficiency: It is said that the energy requirements of chemical processes should be recognized for their environmental and economic impacts and should be minimized. And if there is possibility than synthetic methods should be conducted at ambient temperature and pressure.
- Use of Renewable Feedstock: Rather than depleting whenever technically and economically practicable a raw material or feedstock should be renewable.
- Reduce Derivatives: Unnecessary derivatization (use of blocking groups, protection/deprotection, and temporary modification of physical/chemical processes) should be minimized or avoided if possible, because such steps require additional reagents and can generate waste.
- Catalysis: Catalytic reagents (as selective as possible) are superior to stoichiometric reagents.

- Design for degradation: It has been said that chemical products should be designed in such a way that at the end of their function they break down into innocuous degradation products and do not persist in the environment.
- Real-time analysis for Pollution Prevention: There is requirement for the development of the analytical methodologies which needs to be further developed to allow for real-time, in-process monitoring and control prior to the formation of hazardous substances.
- Inherently Safer Chemistry for Accident Prevention: The substances which are used in the chemical process are the substances and the form of a substance which should be chosen to minimize the potential for chemical accidents, including releases, explosions, and fires.

Five easy things to make the lab more sustainable:

- Close fume hoods when not in use to reduce energy use;
- Run experiments on the microscale to reduce waste;
- Switch to green solvents: Use 2-methyl tetrahydrofuran in place of methylene chloride, and use cyclopentylmethyl ether in place of tetrahydrofuran, 1,4-dioxane and ether;
- Neutralize basic phosphate-buffered HPLC waste or acidic HCl waste to pH 7 and pour down the drain; and
- Recycle electronics, ice packs, packaging materials, toner cartridges, pipette tip boxes, and water purification cartridges.

To reduce the cost green chemistry is spreading from academic labs into industry, as well as environmental, health, and safety risks.

On the large-and small-scale applications of the twelve guiding principles are found on, and it is done only with motive of choosing ingredients for reactions that minimize waste and risk to metrics that quantify waste and process efficiency.

To track energy use during production principles of engineering and process design lead green chemists, and it is done because there is a need to search for sustainable raw materials and build biodegradable or recyclable products to prevent waste.

The elements of achieving sustainable regional production include:

- With good clinical and manufacturing practices coupling indigenous knowledge;
- By virtue of their simplicity identifying technologies that are elegant;
- For advanced technology manufacturing designing a "green footprint;" and
- Utilizing process analytical technologies in a manner that guarantees quality, in addition to rugged, robust manufacturing.

However, green chemistry is as yet not generally executed. Current appraisals put green chemistry items at just 1% of the items from the substance segment. A few hindrances prevent usage of green chemistry in the United States.

The test of creating sustainability measurements shields organizations from assessing their procedures and consolidating green chemistry into business choices. Guidelines around tranquilize creation and the venture tied up in existing synthetic plants prevent the turn of events and execution of new advances.

The interdisciplinary idea of green chemistry additionally challenges the particular information picked up in current preparing and modern work understanding. In the USA, government arrangements that forward information sharing or give financial impetuses can assist prod with greening science development.

In this way, apparently another vision for synthetic training is required, enveloping numerous new measurements, on the off chance that it is to address the difficulties in connecting with natural sustainability.

7.3.2. Sustainable Chemistry Education

It is apparent that sustainable chemistry instruction includes various techniques in showing crucial chemistry ideas, whereby new terms and new methods of reasoning are presented. The center subject of thermodynamics should be talked about as far as energy effectiveness of compound handling and production notwithstanding energetics and immediacy of concoction responses.

The center subject of kinetics should be talked about as far as specific impetuses, which expand item yield by diminishing result arrangement. Such conversations interlink center chemistry information with green chemistry standards and structure the establishment on which sustainability of the concoction undertaking is advanced.

As a result of such incorporations in chemistry educational plans, a set-up of new terms develops, for example, 'feedstock' supplanting 'reactant' and 'E-factor,' which is the proportion of the mass of 'squander' contrasted with that of 'item.' The last is a straightforward observational proportion of the 'green-ness' of a concoction procedure, and subsequently, its sustainability.

So also, a conversation of 'sustainable power source assets' must be introduced by a conversation of present essential energy assets, to be specific petroleum products, so as to address environmental change; contending that these must be supplanted dynamically by perfect, green, sustainable power source assets, for example, sun-based energy.

Sustainable chemistry additionally grasps natural chemistry, whereby principal compound ideas, for example, the p-square components-C, N, O, P, and S-are named 'supplements' and 'salts' are liable for 'saltiness' of soils and surface waters.

Poisons upset the characteristic supplement cycles and saltiness decreases soil and freshwater quality with generally speaking corruption of the regular habitat. Essentially, expanding corrosiveness of waterways and seas upsets sea-going biological systems and is an immediate outcome of expanded degrees of CO_2 in the climate.

Moreover, expanding harmfulness of the earth because of compound waste in soils, air, and surface waters is of most noteworthy worry regarding tending to natural sustainability.

Sustainable chemistry instinctively includes commitment with the age of new savvy materials, and henceforth, with nanotechnology and its visualized linkages to worldwide clean energy prerequisites.

The quickly progressing nano-chemistry is maybe the hugest model of driving edge sustainable chemistry with its emphasis on the improvement of new savvy materials for energy stockpiling, creation, and change, for progressing agrarian profitability, water purging, and desalination food handling, building development, wellbeing checking and for bother control.

Of these applications, fast progression in the creation of photograph voltaic gadgets and carbon nano-tube sunlight-based cells is quickening the sun-based energy industry. Likewise, the advancement of nano-impetuses for hydrogen creation, combined with carbon nano-tube hydrogen stockpiling frameworks is advancing hydrogen as a suitable, elective clean energy asset.

In this way, sustainable chemistry by means of nano-chemistry straightforwardly draws in with ecological sustainability by giving procedures and items which legitimately advantage mankind without hurting nature.

However, these components of sustainable chemistry present impressive difficulties for substance training, both as far as future bearing and extension. Obviously 'sustainable chemistry' can't be considered as a solitary scholastic course yet requires the idea and theory of sustainability to be dynamically brought into all chemistry courses, both at the optional and post-auxiliary/tertiary levels.

Moreover, the unpredictability of sustainable chemistry and the decent variety connected to its usage request adaptable showing techniques, for example, Problem Based Learning upheld with sight and sound grapples, prompting deliberately planned learning results (investigation into which is best case scenario, early stage).

Taking everything into account, since 'sustainability' and 'sustainable turn of events' are mind boggling, multi-dimensional ideas, sustainable chemistry is additionally multi-dimensional in character, grasping orders not ordinarily lined up with it, for example, financial matters, bookkeeping, humanities, social science, social examinations, wellbeing sciences, food science and rural science.

Subsequently, fruitful commitment of concoction training with sustainability includes creating organizations with these controls to frame a unified instructive stage for moving towards ecological sustainability.

In a general sense, sustainable chemistry training is an amazing way of thinking incorporating 'chemistry' into the 'sustainable future' condition and offers provoking instructive chances to accomplish recognizable sustainable results.

7.3.3. Teaching of Green Chemistry

Teaching must be in congruity with training. The topic of how to teach the group of people yet to come of scientists having the aptitude and information to rehearse ecological chemistry lies in the focal point of instructive materials identified with green chemistry.

Training is particularly significant in the advancement of green chemistry. Distinctive worldwide organizations, for example, the American Chemical Society (ACS) and Polish Chemical Society (PTChem), are dynamic in distributing materials that advance the standards and accomplishments of green chemistry.The green chemistry program should prompt sustainability by structuring and utilizing the strategies where common crude materials will be financially prepared, reasonable utilization of energy sources, end of unsafe vaporous, liquid, and strong squanders and by presentation of security items for man.The advancement of green chemistry in schools, among the laborers at plants of chemical industry and wholesalers of chemical items is additionally significant.The expansive use of green chemistry accomplishments will empower us to adjust eco-advancement beneficial for society, economy, and nature. The various instructive materials, accessible at present on showcase and on the Internet, are helpful in regular educating of green chemistry standards.

7.4. PRINCIPLES AND APPLICATIONS OF GREEN CHEMISTRY

7.4.1. Prevention

It is smarter to remove waste than to treat or tidy up waste after it is framed. The capacity of scientific experts to overhaul chemical changes to limit the age of unsafe waste is a significant initial phase in contamination avoidance. It returns to the familiar axiom "avoidance is superior to fix." It is smarter to forestall squander than tidy it up sometime later.

7.4.2. Atom Economy

This guideline gets into the genuine chemistry of how items are made. This standard express that it is ideal to utilize all the particles in a procedure. Furthermore, those molecules that is not utilized end up as waste. Picking changes that join the vast majority of the beginning materials into the item are increasingly productive and limit waste.

7.4.3. Less Hazardous Chemical Synthesis

The objective is to lessen the danger of the chemicals that are utilized to make an item. Scientific experts have generally utilized whatever implies fundamental. Today we are finding that less unsafe reagents and chemicals can be utilized in a procedure to make products.Synthetic techniques ought to be intended to utilize and create substances that have practically no poisonousness to human wellbeing and condition. Some harmful chemicals are substituted by more secure ones for a green innovation.

For instance, in the assembling of polystyrene froth sheet pressing material, CFCs which add to O_3 exhaustion and a worldwide temperature alteration, have now been supplanted by CO_2.

7.4.4. Designing Safer Chemicals

Everybody needs safe items. This rule is planned for structuring items that are protected and non-poisonous. Pharmaceutical items regularly comprise of chiral atoms and the contrast between the two structures can involve life and demise, for instance, racemic Thalidomide when controlled during pregnancy, prompts loathsome birth abandons in numerous new conceived.

Proof shows that just one of the enantiomers has a relieving impact while the other isomer is the reason for extreme imperfections (Figure 7.3).

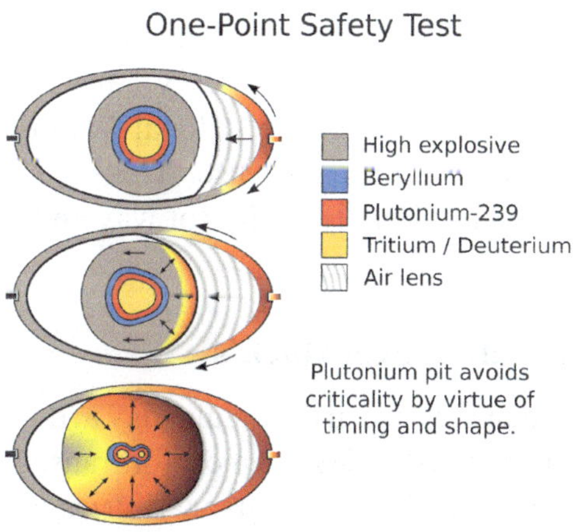

Figure 7.3. Safer chemical test.

Source: Image by Wikipedia.

7.4.5. Safer Solvents

We use solvents normally in our everyday lives (cleaning items, nail polish, beautifiers, and so forth.) and in the chemistry lab, numerous chemical responses are done in a dissolvable. What's more, customarily natural solvents have been utilized that posture risks and many are exceptionally harmful.Solvents are widely utilized in the vast majority of the unions. Broadly utilized solvents in combinations are poisonous and unstable liquor, benzene (known cancer-causing), CCl_4, $CHCl_3$, perchloroethylene, CH_2Cl_2.

Sanitization additionally uses a lot of solvents (for example Chromatography) which add to contamination and can be exceptionally perilous to human wellbeing.

The improvement of Green Chemistry reclassifies the job of a dissolvable: A perfect alluring green dissolvable ought to be regular, nontoxic, modest, and promptly accessible. This rule centers on making items so that they utilize less unsafe solvents. Clearly, water is the most reasonable and naturally considerate dissolvable.

7.4.6. Design for Energy Efficiency

Today there is an emphasis on sustainable power source and energy protection. We use energy for transportation purposes and to give power to our homes and organizations.

Customary techniques for creating energy have been found to add to worldwide natural issues, for example, Global Warming and the energy utilized can likewise be a critical expense. This rule centers on making items and materials in a profoundly proficient way and diminishing related contamination and cost.

7.4.7. Use of Renewable Feedstock's

90–95% of the items that are used in the regular day to day work are produced using oil. The society not just relies upon oil for transportation and energy, yet in addition for making items. This standard tries to move our reliance on oil and to make items from sustainable materials.Biodiesel is one case of this where analysts are attempting to discover elective fills that can be utilized for transportation. Another model is bio-based plastics.

Polylactic corrosive (PLA) is one plastic that is being produced using inexhaustible feedstocks, for example, corn, and potato waste.

Benzene utilized in the business amalgamation of adipic corrosive which is required in the production of nylon, plasticizers, and ointments, has been supplanted somewhat by the inexhaustible and nontoxic glucose and the response is done in water.

7.4.8. Catalysis

In order to make reactions happen more efficiently and also to reduce energy requirements catalysts are basically used in a chemical process. Generally, to have a required effect on the reaction only a small amount of catalyst is required it is the benefit of a catalyst.

Further, any kind of toxicity is not observed in the process if the catalyst used in the process is purely a green catalyst as it is the property of green catalysts. The most wonderful examples of catalysts are enzymes. As bio catalyzed reactions are performed in an aqueous medium so these reactions are advantageous.

7.4.9. Design for Degradation

The persistence of the renewable resources within the environment and also there is a need to obtain the materials and the products from these renewable resources. Further, the products used in day-to-day life are persistent and no question could rise related to this.

In the water streams pharmaceutical drugs as antibiotics are building up in it, also plastics do not degrade in the landfills. To perform their intended function efficiently these products are seeking to design in such a way.

7.4.10. Pollution Prevention

Pollution control is better than this pollution as the fact that prevention is always better than the cure similarly in this case it works. To eliminate pollution or the wastes at the sources, there are some of the process and practices and some materials which help in pollution control.

7.4.11. Safer Chemistry for Accident Prevention

Where an industry resides their safety of the surroundings and the workers should be ensured, and their principle should entirely focus on these aspects. Light on fire, ignite in air, etc., are some of the materials which will not explode so; these materials are preferred to use as they are comparatively safe to use in order to make a product.

In some areas, the results could be disastrous where safe chemicals were not used. One such example is Bhopal gas tragedy in the year 1984 and this tragedy was the most widely known as perhaps the most devastating disaster.

Thousands of lives were lost and many more were injured due to the chemical plant had an accident release. The reactive chemical which can prove to result in the accidents these chemicals should be avoided in the creation of products. The results are often devastating in case of explosion and fires happen.

7.5. GREEN CHEMISTRY SKILLS

Under the new Green chemistry concepts in order to develop new processes, synthetic concepts, reaction conditions, catalysts, etc., chemists from all over the world are using creative and innovative skills.

The novel academic researchers have been led to examine alternatives to the existing synthetic methods. These novel academic researchers have been led by the commercial applications of green chemistry.

Some of these applications are:

- The synthesis of polycarbonates has been replaced by diphenyl carbonate and this polycarbonate has been synthesized by the use of phosgene and methyl chloride.
- Oxidation in the industry is the most polluting reaction. The use of alternatives to fewer polluting reagents is basically led by the implementation of green chemistry. Further by using molecular O_2 as the primary oxidant, the metal ion contamination is minimized. And also, to minimize the use of extremely high oxidation state transition metal complexes.
- By Wacker a convenient green synthesis of acetaldehyde, in the presence of a catalyst the oxidation of ethylene is done with O_2. H_2SO_4 is used in hydration of acetylene or oxidation of ethanol.
- The solubility of the metal ion is basically increased by the chelates which are the complex substances and they basically interact with the metal ions. In many types of industrial process and cleaners these chelates are being used.

Environmental hazards are caused by the use of conventional methylation reactions these in turn leads to employment of toxic alkyl halides or methylsulfate. To avoid environmental hazards are avoided by replacing these toxic substances by the means of dimethyl carbonate and also inorganic salts are also not deposited.

For 100% CO_2 blowing agent for polystyrene, Greener Reaction award was won by the Dow Chemical in year 1996. In packing and food transportation the common material which is used is termed as the polystyrene foam.

In the process of production of the foam sheets which present serious environmental hazards are traditionally being produced by the means of the CFC and other ozone depleting chemicals. Without the need for hazardous substances the supercritical CO_2 works as equally as well as a blowing agent and this has been discovered by the Dow Chemical.

This CO_2 also allows the polystyrene to be more easily recycled. The net carbon which is released from the process is zero as the CO_2 used in the process is reused from other industries. Including detergents, polyurethanes, and food additives and other variety of products are obtained by the means of propylene oxide (PO) which acts a chemical building block for their production.

The CO products such as t-butyl alcohol, styrene monomer or cumene are traditionally led by the chlorohydrin which is led by PO production. Including a significant amount of waste, the by-products created by the manufactures.

Propylene oxide mainly eliminates out most of the waste. This PO with hydrogen peroxide is basically developed a new route jointly by the Dow and BASF.

In refrigerators and air conditioners traditionally the CFCs have been used. These CFCs act as refrigerants in them. They have a disadvantage towards the environment they led to the depletion of the ozone layer.

Apart from the disadvantage, it has some of the advantages such as safe incombustibility, high stability, and low toxicity. CFCs replace hydrofluorocarbons (HFCs) and hydrochlorofluorocarbons (HCFCs) in the past decades. Further, it is observed that these HCFCs and HFCs are indeed safer in the case of the ozone layer. They do not lead to ozone depletion.

Phosphates example sodium tripolyphosphate and amino carboxylic acids such as ethylenediaminetetraacetic acid (EDTA) are the ones on which these conventional chelates are based upon.

These conventional materials are observed as environmentally unfriendly because phosphates can basically cause pollution by the means of eutrophication and also unfortunately EDTA is not readily biodegradable. The biodegradable chelating agent which is principally manufactured from a renewable fed stock has been developed by Akzo Nobel.

7.6. GREEN CHEMISTRY AND THE INNOVATION ENGINE

In history, it has been seen, the tactics and techniques that have been used in order to cope up or face the issues which were related to the environment, more particularly, which were related to the chemicals and materials have been in order to decrease with the help of decreasing the exposure.

On the other hand, this made sense at that point of time, and the reason behind this is the fact that there were lack of understanding of the basis of the toxicity and hazard.

If the level of awareness of a hazard is simply that it will produce a contrary result, then the only option or only alternative is to avoid the hazard where the avoidance, with respect to the time meant lessening the exposure or any kind of contact to it (Figure 7.4).

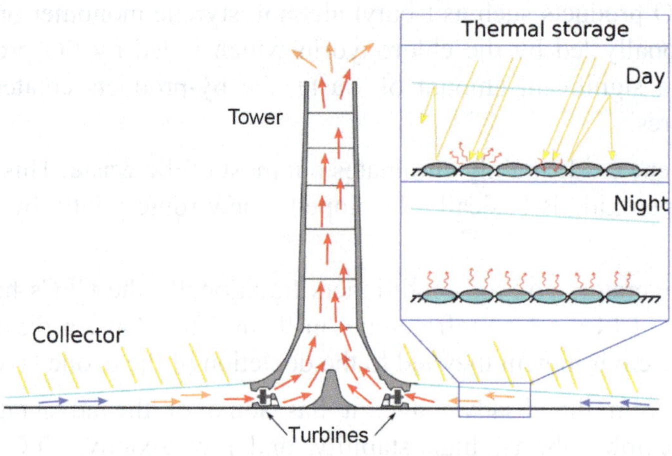

Figure 7.4. Innovation engine and the green chemistry.

Source: Image by Wikipedia.

The methodologies that were established within this model were every so often, very complicated with respect to the technicalities and also, sometimes clever, or resourceful ways of working with hazards and reducing any kind of contact with them.

All of these approaches had two things in common, as mentioned below:

1. They could (and as a probability function would) fail; and
2. All of these approaches essentially had to be a cost drain.

As elegant as several numbers of these technological bandages were, they did not add any kind of value to the product or process with the help of which they were linked.

As people have attained an increment in the level of awareness, with respect to the fundamental or actual nature of hazard, it has provided the ability in order to cope up with or face the hazard at the stage of design.

With the acknowledgment of the molecules, the chemist designers can theoretically develop molecules as well as the molecular transformation, in order to decrease or remove the chemical as well as physical characteristics of the hazard. This theory is extensively used in practice, within the field of Green chemistry.

The ability plays an important role in the form of a tool for sustainability, with respect to the power of Green Chemistry. Not likely the exposure controls that have been created in history.

Green Chemistry can convey the added performance, added effectiveness, and added capabilities to the products and processes it is applied to, at the samc point of time that it faces the issues that were related to the environment, along with the human health.

7.7. BIOCATALYSTS

It is extensively understood that the biocatalysts along with the enzymes have several numbers of various benefits; on the other hand, both of them have some major disadvantages as well.

According to the latest authoritative text, Bommarius and Riebel, with respect to the enzymes, the major or important things to remember is that their extremely good selectivity has to be weighed against their lack of stability. Apart from this, there are several other general considerations as well.

First and foremost, according to the Klibanov's work, where it was depicted that several numbers of various enzyme can function perfectly satisfactorily with respect to the non-aqueous solvents, generally, enzymes are far better in their general performance in aqueous media.

Apart from that, the second is, some of the chemical reactions cannot be catalyzed with the help of the enzymes. Specifically, this is true for those, who act as a central role in the industrial manufacturing.

Third and the last is, usually, enzymes necessitate the participation of expensive co-factors (for example, NADPH). In an ideal condition, it would be better to have just one important or meaningful catalytic agent (indeed, this can be done, or this can be accomplished in some of the enzymes bred by a well-known researcher named, Arnold, as well as others' inorganic enzyme mimics).

The prerequisite for a co-factor is quite restraining, and the reason behind this is that, it prevents the application of the enzymes in vitro without the addition of a coupled, co-factor-regenerating system.

7.7.1. Enzymes

Quite apart from the subtle (and comprehensive) alterations or modifications that can be made, with respect to the sequence of amino acid is more or less protein with the help of the process that is known as site-selected mutagenesis (Figure 7.5).

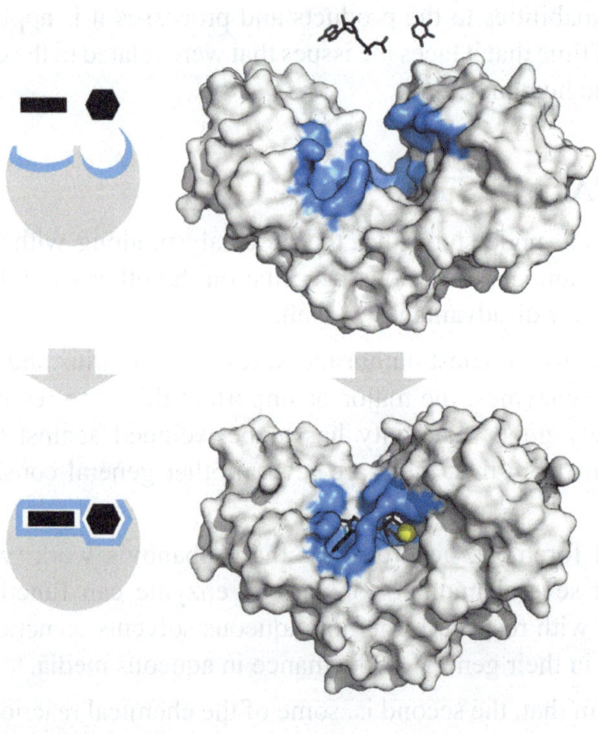

Figure 7.5. Enzymes.

Source: Image by Wikipedia.

In addition to this, also, there are other methods or other techniques of developing the powerful enzyme catalysts, and maybe the most significant of these is that of Darwinian evolution.

In effect, the philosophy or the ideology behind this technique is based on the fact that, as an individual does not really know what series of amino acids will confer the needed catalytic activity and the selectivity, it is best, in spite, in order to breed the proteins (with the help of forced evolution), in order to accomplish the novel and desirable functions.

The beginning point of the Reetz's biocatalysis is a kind of wild-type enzyme that increases the rate of the reaction or catalyzes a reaction of interest, but not enantio-selectivity. An instance is the lipase from the bacterium Pseudomonas aeruginosa that catalyzes the hydrolysis of esters.

The genes of that encode the enzyme is first subjected to the random mutagenesis in order to produce a library of mutant genes in a test tube. The library is then inserted into an appropriate micro-organism and articulated as a library of the mutant enzymes, which are screened for enantio-selectivity.

Mediocre mutants are rejected, and the mutant gene of the enzyme depicting the optimal enantio-selectivity is then subjected for the further cycles of the mutagenesis, expression, as well as screening.

As near about two thousand to five thousand mutant enzymes are developed, it is very important in order to used high-throughput screening for the developed enantio-selectivity (Reetz has developed Nuclear Magnetic Resonance (NMR) and mass-spectroscopic methodologies or tactics in order to allow them in order to identify up to ten thousand enantiomeric excess (EE) values per day.

Taking an instance, Reetz et al. have vividly improved the enantio-selectively of a lipase in the hydrolytic kinetic resolution of a chiral ester. They enhanced the selectively factor from 1.1 to more than fifty-one.

Moreover, in order to increase the rate of the reaction of hydrolysis or catalyzing the hydrolysis mentioned above, Reetz et al. produced, with the help of directed Darwinian evolution, an enantio-selective cyclo-hexanone mono-oxygenase for the Baeyer-Villiger oxidation of 4-hydroxy cyclo hexanone in order to form two lactone enantiomers.

7.7.2. Solid Inorganic Enzyme

Metallo-enzymes having a metal active site owe their high selectivity in the chemical reactions mainly to the tertiary structure of protein.

This structure wraps around the active site and performs various kinds of important and necessary operations, for example, safeguarding the active site from the harmful or toxic reactions with the help of the biomolecular pathways, separating or filtering the substrate molecules.

It is done in such a way that, only the molecules that are having the potential or capable of passing across the protein channels can attain the access to the embedded prosthetic group, and offering a stereochemically demanding empty space, in the vicinity of the active site, where the substrate residues through the course of the catalytic reaction (Figure 7.6).

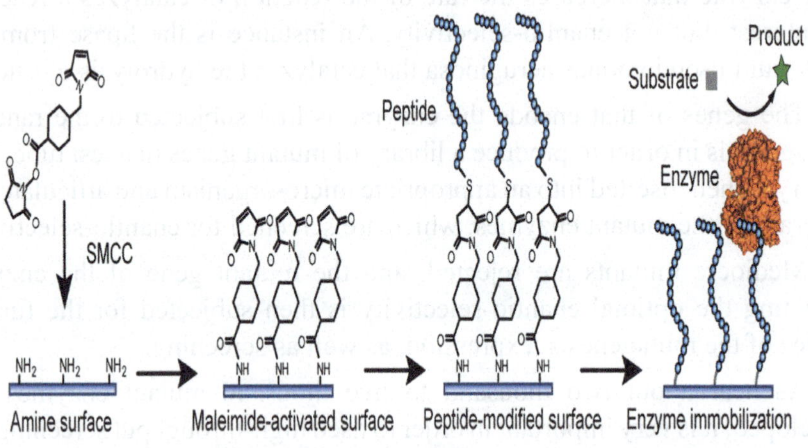

Figure 7.6. Industrial enzymes.

Source: Image by Wikipedia.

The open framework of micro-porous as well as the other zeolitic solids has several numbers of structural resemblances extraordinarily identical to the protein mantle of the natural enzymes. They are consisting of pores, voids, and the channels of molecular dimensions that optimum only those kinds of molecules that are of appropriate size and shape in order to pass across the pore.

In addition to this, they also orient those molecules (that are based on their physical dimensions) that gain access to the internal empty spaces of the crystallite, where the active site is situated.

With the help of taking benefits of these kinds of the resemblances, a new catalyst that integrates the important as well as vital characteristics of the robust, chemically inert molecules filter with the high activity and the selectivity of the natural enzymes can be developed.

In addition to this, at present interval of time, as much is known about the methodologies or tactics of designing the molecular dimensions, degree of hydro-phobicity, along with the other characteristics of the open-structure inorganic catalysts, genuine inorganic mimics of the specific kinds of enzymes can be prepared.

Moreover, a variety of highly refined methods has been changed or grew to probe, under operating settings, the behavior of the species of the reactant, as well as the active sites that are present inside the cavities of the open-structure inorganic solids.

Herron, Raja and Ratnasamy, Jacobs et al., Bein et al. and Backvall et al. have designed zeolite-encapsulated metal complexes that are made up with the help of porphyrins, phthalocyanines, salens, salophens, bipyridines, and tri methyl triazo cyclononane (TMTACN) for several number of oxidation reactions.These kinds of site-encapsulated catalysts permits access of only those reactant molecules that are enough small in size, so that they can diffuse through the windows of the zeolitic cage and avoid any kinds of these products, which, if created are too big in size, from leaving the sphere of the reaction.

More necessarily, the encapsulation or the cocoon safeguards the entity of the active site from losing its intrinsic catalytic performance because of the dimerization or other complicating side-reactions.

These kinds of catalysts, every so often, known as zeo-zymes or enzyme mimics for obvious reasons, have proved efficient in oxy-functionalizing the terminal methyl groups of linear alkanes, in order to transforming the CH_4 into methanol and propane into iso-propanol, with respect to the epoxidizing alkenes with H_2O_2 with high productivity and selectivity, and for the aerobic oxidation of the alcohols.

Separately, from oxidizing the typical benefits of being able to eradicate and reuse the ship-in-bottle catalyst with the help of process of simple filtration, these workers displayed that these kinds of single-site heterogeneous catalysts had a higher specific rate, when they are compared with their homogeneous analogs.

7.7.3. Acid-Catalyzed Reactions

For several numbers of decades, acid catalysts have been employed on a commercial scale, but the requirement or the necessity to develop new ones has recently increased in order to reduce the disadvantages that are linked with the present extensive application of corrosive and environmentally hazardous liquids, for example, concentrated phosphoric, sulfuric, and hydrofluoric acids as well.The acids that are considered as very strong acids are having the potential to catalyze the "demanding" reactions, for example, alkylation (for instance, catalyzation of iso-butene with the help of butane) in order to produce octane fuels or the isomerization of the C_5 to C_9 n-alkanes, for example, liquids mentioned above, are every so often supported on the siliceous or clay minerals

But also, there is an urgent requirement in order to create new solid catalysts. That kind of catalysts is that do not suffer from the environmental disadvantages and opposing the corrosive characteristics of these kinds of liquids.

In addition to this, also, there is a requirement for new, intrinsically weaker, acidic catalysts that would be very beneficial with respect to the production of ethers and small hydrocarbons with the help of the dehydration of alkanols under rather mild settings.

With respect to this, a short discussion of these kinds of tactics or methods has been given elsewhere. In all so-called pentasil zeolites (for example, ZSM-5 (zeolite socony mobil-5) and ZSM-11), where the ratios of silicon to aluminum (Si / Al) are seldom less than 10, and in several numbers of other aluminosilicate and alumino-phosphate open-structure solids (for example, faujasite, ferrierite, erionite, and SAPO-34), a large variation of (Bronsted) acid-catalyzed reactions may be carried out.

Apart from this, one very important acid-catalyzed reaction merits further comment. It is the dehydration of the methanol in to (mainly) ethylene and propylene (with traces of the butanes). These are the feedstocks for the polyolefin market, which is huge and massive.

The production of the poly-ethylene is anticipated to elevate from 55 $\times 10^6$ metric tons in the year of 2002 to near about 87 $\times 10^6$ metric tons in the year of 2010. On the other hand, the production of the polypropylene is anticipated to elevate from 35 $\times 10^6$ to 60 $\times 10^6$ metric tons in the same interval of time.

One of us has been involved in the designing of the shape-selective solid acid catalysts, which are very effective with respect to the production of the light olefins (which is mainly ethylene and propylene) in the dehydration of the methanol. There are two significant instances, MAlPO-18 (M ≡ Zn ^{II}, Mg ^{II}, Mn ^{II}, Co ^{II}) and Davy Faraday microporous solid No. 4 (DAF-4).

7.8. FOSTERING INNOVATION

The combination of Green Chemistry into early product development will offer improved value with the help of the product life cycle.

In order to mitigate the double economic consequences which are linked with the generation of waste the purchase of chemicals that are not converted into the product, along with the cost to dispose of this waste, the Twelve Principle should be applied in early stages of the product development.

Green Chemistry can convey the added performance, added efficacy, as well as added capabilities to the products and processes it is applied to the same interval of time that it addresses the concerns or issues that are related to the environment as well as human health.

Metrics, which is particularly used in the form of enhancing life cycle tools and developing collaboration in order to make their application easier, required being a focal point in order to move forward.

People have to understand and acknowledge how to find the solutions to the hurdles of the information that is transferred from supplier to the manufacturer to the end-user and in order to determine the mechanisms to control or organize the intensity of the data of the process to make sure that the life cycle metrics are developed and executed.

Never the less, there is a clear requirement for improved use of metrics, it is the direction and magnitude of the change that is necessary. The Presidential Green Chemistry Challenge Award program has shown the success of Green Chemistry and the effect innovation can have on the environment and on the corporation's bottom line.

7.9. CONCLUSION

Green Chemistry is the design, development, as well as the execution of the chemical products and processes in order to decrease or remove the application and generation of substances that are dangerous to the environment as well as dangerous to the human health.

The integration of the Green Chemistry into the field of science curricula will aid the drive for innovation and sustainability. The Twelve Principles of

Green Chemistry offer the framework to the sustainability at the most basic level.

In the present interval of time, the principles of Green Chemistry are required to become the core for tomorrow's chemistry, with the help of combining sustainability with the science and its innovations.

Different from the regulatory necessities for the pollution prevention, Green Chemistry is an innovative, non-regulatory, as well as an economically driven approach in the direction towards the sustainability. Green Chemistry contemplates the complete life cycle of the chemical processes as a chance for design innovation.

In spite of the regulatory restrictions for controlling dangers and threats, Green Chemistry challenges innovators in order to design and exploit matter as well as energy in such a way that it elevates the performance and value, through the course of safeguarding the human health along with the environment.

With the help of a discussion of latest innovations as well as collaborations, the progress of Green Chemistry will be demonstrated, and difficulties will be presented. With a concentration on the innovation, there is always a chance for the regular or constant enhancement.

It is the frontiers of Green Chemistry which provides the challenges and, side-by-side, the momentum, in order to design for sustainability.

REFERENCES

1. Dhage, S. D., & Komalsing, K. S., (2013). Applications of green chemistry in sustainable development. International Research Journal of Pharmacy, 4, 1–4.
2. Manley, J. B., Anastas, P. T., & Cue, Jr. B.W., (2008). Frontiers in green chemistry: Meeting the grand challenges for sustainability in R&D and manufacturing. Journal of Cleaner Production, 16(6), 743–750.
3. Meurig, T. J., & Raja, R., (2005). Designing catalysts for clean technology, green chemistry, and sustainable development. Annu. Rev. Mater. Res., 35, 315–350.
4. Sambhudevan, S., (2018). Green Chemistry and Sustainability | Amrita Vishwa Vidyapeetham. [Online] Amrita.edu. Available at: https://www.amrita.edu/news/green-chemistry-and-sustainability (accessed on 4 July 2020).
5. Song, J., & Han, B., (2015). Green chemistry: a tool for the sustainable development of the chemical industry. National Science Review, 2(3), 255–256.

Chapter

8

Global Perspective on Energy and Sustainable Development

CONTENTS

In the chapter, a global perspective on energy and sustainable development, various steps undertaken by government of different countries to attain sustainable development goals for energy is discussed. The chapter discusses the sustainable energy in detail and various approaches to energy sustainability are also explained. In addition, several drivers for renewable energy policies are also discussed.

Some of the primary ones are environmental drivers, economic drivers, and political and security drivers. Major policy steps for energy for sustainable development are also discussed. Various policy options for promoting technological innovation are also being explained. Several challenges faced by the experts in achieving energy sustainable development goals are discussed. It also explains the role of the World Bank. In the end, it explains the case of community energy.

8.1. INTRODUCTION

Societies around the world are on the verge of a profound and instantly needed transformation in the way they generate and consume energy. This dramatic shift is moving the world away from the consumption of fossil fuels (which is the preliminary cause of climate change and also a major contributor to other social and environmental challenges) toward cleaner, renewable forms of energy.

Around one-fourth of the population all across the world are using renewable energy significantly in order to generate electricity, cool, and heat buildings, cook, and provide mobility.

Renewable energy is price competitive and market-ready with conservative sources in many jurisdictions and met about 19% of the world's final energy demand in 2014.

Around the world, islands, communities, and cities have found that making the transition to 100% renewable energy is typically a stuff of political will and that the needed technologies are already present in the market.

Government from various countries all around the world and on all continents is setting impressive targets for renewable energy, with an ever-growing number of jurisdictions aspiring for 100% of renewables.

Local governments, in particular, are backing up this movement and are becoming incubators of locally suitable best policies and practices.

The noteworthy engagement of renewable energy has been driven primarily by a variety of objectives (drivers) that incorporate advancing economic growth, augmenting energy access, comprehensive energy security, and alleviating climate change.

Altogether, these drivers might be defined as the quest of sustainable development, where economic affluence is advanced across the world while negative impacts are minimized.

While such notable advantages are extensively cited as key drivers in energy and political debates, specific, documented evidence of such advantages remains rather restricted for reasons involving a lack of acceptable conceptual frameworks, methodological confronts, and partial access to relevant data.

8.2. ENERGY SUSTAINABILITY

Figure 8.1. Cleaner energy.

Sources: Pixabay.

In some ways, the energy sustainability concept is generally the application of the typical definitions of sustainability to energy. In other ways, energy sustainability is more intricate and involved (Figure 8.1).

Energy sustainability basically refers to the inclusion of energy services in a sustainable manner, which in turn demands that energy services be provided for all people in ways that, in present as well as in the future, are adequate to provide basic necessities, affordable, acceptable to people and communities, and having no negative impact to the environment.

There has been not, yet any universal definition of energy sustainability defined, but various descriptions and definitions have been presented. Several connections between energy sustainability and factors such as economic growth and efficiency have been investigated. The relationship between energy, sustainable development and the environment is worth considering. The supply of energy as well as their use is associated with climate change as well as such environmental concerns as ozone depletion, air pollution, emissions of radioactive substances, and forest destruction.

All these issues need to be addressed for the overall benefit of society, while at the same time maintaining a clean and healthy environment, especially because the future will be at stake due to the rate at which societies and people are degrading the environment.

A society looking for sustainable development need to utilize only those energy resources that have no potential impact to the environment (for instance, that release no emissions or only emissions which are not harmless to the environment).However, since all energy resources have some kind of impact to the environment, environmental stewardship and, improved efficiency and effectiveness can help in overcoming many these concerns in context to the restriction enacted on sustainable development by environmental emissions and their consequential negative impacts.

8.3. APPROACH TO ENERGY SUSTAINABILITY

There are numerous individual components to the manner in which energy resources can be used in sustainable way in society, each of which is a pre-requisite for energy sustainability. Various aspects of energy sustainability are described and examined is defined and explained in the next following sections:

8.3.1. Harness Sustainable Energy Sources

The necessities for energy services need to be fulfilled, and sustainable energy sources need to be employed for that goal. This will be a comparatively difficult task as the use of energy resources increases with high living standards and increasing populations, particularly as developing countries become more affluent and industrialized. Many renewable as well as non-renewable energy sources are explained in Tables 8.1 and 8.2, correspondingly. Non-renewable energy sources generally comprise energy resources that can be available in restricted quantities.

Table 8.1. Non-Renewable Energy Sources

Fossil Fuels	Non-fossil fuels
Coal	Biomass (when not replenished)
Oil	Uranium
Natural gas	Fusion materials (e.g., deuterium)
Tar Sands	Wastes
Oil Shales	
Peat	

Source: Mdpi.com

Table 8.2. Renewable Energy Sources

Solar related energy	Non-solar related energy
Hydraulic energy	Geothermal energy (ambient)
Wave energy	Geothermal energy (hot)
Wind energy	Tidal energy
Ocean thermal energy	
Biomass (when replenished)	

Source: Mdpi.com

Fossil fuels are non-renewable energy resource, as are other sources such as uranium, although nuclear fuel lifetimes primarily relied upon the advanced nuclear technologies such as breeder reactors. Renewable energy comprises solar radiation incident on the earth, and energy forms that is conducive to radiation, as well as energy from such other natural forces as the rotation and gravitation of the earth.

Solar radiation, which is present on the Earth at almost 20,000 times the global energy-use rate, can be gathered as heat or concentrated to high-temperature heat or directly transformed into electricity and stored in photovoltaic (PV) devices.There are various other renewable energy sorts that can be stem from solar radiation. Hydraulic energy, which includes running and falling water, varies in size from huge complexes like at Niagara Falls to much smaller systems such as micro hydro. Ocean thermal energy is basically dependent on the temperature difference between surface and deep waters, which can be manipulated to produce electricity at large.

There are basically two discrete types of geothermal energy: hot, which employs the internal heat of the Earth, and ambient (i.e., natural ground temperature), that can be employed by ground-source energy systems such as heat pumps.

Tidal energy manipulates the motion with tides triggered by gravitational forces of the moon and sun and the revolution of the earth, while wave energy is dependent on the motion of waves.

Biomass energy comprises wood (e.g., forestry wastes or fast-growing trees), plants, and other organic matter, which can be transformed or combusted to other fuels. Biomass energy refers to only a renewable resource till when its total usage does not exceed its total demand.

Wastes (heat and material) can be considered either non-renewable or renewable energy and can be used to generate electricity or produce heat via waste-to-energy incineration.

Non-fossil fuel energy options eradicate or reduce greenhouse gas (GHG) emissions and thus often encourage using sustainable energy solutions, although some such as biomass can produce net emissions if not managed or controlled in an effective and efficient manner.

Although it is generally seen that non-renewable, nuclear energy evades GHG emissions. Nonfossil fuel energy options are diverse with respect to their types, and characteristics, ranging from renewables to nuclear energy.

8.3.2. Utilize Sustainable Energy Carriers

Figure 8.2. Use of renewable energy: Sustainable energy.

Source: Wallpaper Flare.

It is generally seen that sustainable energy carriers usually implies the conversion of sustainable energy sources into appropriate energy carriers. The range of energy carriers is diverse (see Tables 8.3 and 8.4) (Figure 8.2).

Material energy carriers basically comprise secondary chemical fuels, vacillating from conventional ones such as oil products (e.g., diesel fuel, gasoline, naphtha), coal products (e.g., coke) and synthetic gaseous fuels (e.g., coal gasification products), to non-conventional chemical fuels such as methanol, hydrogen, and ammonia.

Thermal energy can be either heat (or a heated medium such as steam, hot air, and exhaust gases) or cold (or a cooled medium such as ice and cold brine) and can be conveyed through district energy systems.

Table 8.3. Material Energy Carriers

Fossil Fuels (Upgraded)	Other Chemical Fuels
Oil products (e.g., gasoline, diesel fuel, naphtha)	Hydrogen
Synthetic gaseous fuels (e.g., coal gasification products)	Methanol
Coal products (e.g., coke)	Ammonia

Source: Mdpi.com

Table 8.4. Non-Material Energy Carriers

Work
Electrical energy
Thermal energy Heat (or a heated medium) Cold (or a cooled medium)

Source: Mdpi.com

For instance, buildings in some cities are coupled with pipes through which steam or hot water flows to provide water and space heating, while on the other hand, district cooling uses a piping network in order to provide cooling.Energy carriers in a broader way are a noteworthy consideration in energy sustainability because conventional and non-fossil fuel energy options are not adequate enough to avoid environmental issues such as climate change, because of the fact that can be utilized instantly in their natural forms. Conversion systems are often required to render non-fossil energy more expediently utilizable.

Hydrogen energy systems are comparatively more essential as they facilitate the use of non-fossil fuels by letting them to be transformed into two main classes of energy carriers: electricity and hydrogen. Together, these two energy carriers provide the path to fulfill most chemical and non-chemical energy needs in a sustainable manner, as long as carriers are generated from sustainable energy resources.

The demand of hydrogen as an energy carrier is anticipated to surge intensely over the next few decades, as a "hydrogen economy" emerges. In such an energy economy, hydrogen will be probable to use in transportation, power generation, and oil sands processing. Several approaches and strategies to a hydrogen economy have been reported by nations, e.g., the United States, Canada, and Iceland.

One critical element of a hydrogen economy is hydrogen production. It is important to note that hydrogen can be extracted from fossil fuels, as well as processes not potentially fossil fuel-based, incorporating water electrolysis using electricity and thermochemical water decomposition via electrical and/or thermal energy.

The latter two processes split water into oxygen and hydrogen. There is need to examine various pathways to hydrogen energy, concentrating on options of renewable energy and the potential contributions of nuclear energy.

8.3.3. Increase Efficiency

It is generally seen that high efficiency allows the maximum benefits that can be attained from energy options and reducing potential environmental impact, and thus supports sustainability initiative. Efficiency improvements also comprise fuel substitution, energy conservation, better coordination of energy carriers and demands, efficient energy management, and more effectual consumption of energy quality.

Of course, the overall effectiveness of energy efficiency measures must be measured, and many have pointed out that, contrary to expectancies, smarter way of energy use occasionally through "rebound" effects resulted in meager or no fall in the utilization of energy resources, and sometimes led to high consumption.

This kind of rebound effects have triggered some to object the degree in which efficiency can enhance quality of environment and encourage sustainability.

In addressing efficiency of energy use, it is important to note that that energy is portrayed by the laws of thermodynamics, with the first law exemplifying the principle of energy conservation and the second law concerned with the quality of energy and often including exergy and entropy concepts. Many proficiency amendments are best considered with exergy, which stems from the second law of thermodynamics.

Exergy is almost related to energy in some forms, but dissimilar in others, e.g., the exergy of a substance or an energy form is a measure of its quality or usefulness. Exergy can be defined as the maximum amount of work that can be produced by a flow of energy or matter as it comes to equilibrium in context to the environment, and is a gauge of the potential of a flow to cause change as a consequence of being in disequilibrium with the reference environment.

8.3.4. Reduce Environmental Impact

It is important to note that several environmental impacts concerned with the energy processes are of high importance and should be addressed in efforts to accomplish energy sustainability, including:

- Acidification (refers to overall effect on soil and water of acidic emissions);
- Ozone depletion (for instance, devastation of the atmospheric ozone layer and successive surges in ultraviolet coming in contact of the earth's surface);
- Abiotic resource depletion (because of extraction of non-renewable raw materials);
- Radiological effects (e.g., radiogenic cancer morbidity or mortality due to external or internal radiation exposure);
- Ecotoxicity (health problems due to high exposure to toxic substances);
- Global warming (accredited primarily to GHG emissions and considered to be a critical element in climate change); and
- Improve Socioeconomic Acceptability.

Many socioeconomic sustainability factors associated with energy should be taking into consideration while designing approaches and strategies to energy sustainability.

These factors often are interrelated with the formerly defined sustainability approaches, e.g., exploiting sustainable energy sources must account for global stability, economics, and geographic and intergenerational equity to preserve a sustainability focus.

Various discussion on energy systems have been undertaken by considering them as socio-technical arrangements with strong correlation between social and technological elements such as regulations, institutions, social practices, cultural values, as well as interests, expectations, and relationships of the actors involved.Others have examined socio-technical factors by pondering energy transitions to sustainability and scrutinized the experiences with such transitions of countries that have initiated policies of energy transition policies, for instance, the Netherlands.

This research basically aimed at identifying difficulties to energy transitions, different routes a transition can take and the policy and institutional settings that upkeep such transitions. Some energy-related socioeconomic sustainability factors are discussed as follows:

- Community involvement and social acceptability. People and communities should actively take part in the energy-related decisions in order to achieve the energy sustainability goals, as the support of these groups is essential to triumph of any initiatives, and such support almost always necessitate consultation and participation in decision making. A culture of sustainability can be evolved when a collaborative and consultative approach is steadily followed.
- Economic affordability and equity. In order to attain the sustainable goals, energy services concerned with fulfilling basic needs must be economically affordable by all people, various groups, and societies.

It is worth noticing that some energy initiatives are affordable at present, e.g., some efficiency enhancement and environmental mitigation methods can be implemented in ways that save money and efforts over time. It is important on part of all societies to be able to access energy resources, irrespective of their geographic location, to achieve energy sustainability.

These ideas demand the effort of equity and coordination among developing and developed nations in terms of energy opportunities.

Sustainable energy options should take into notice population growth rate as it emphasizes strains on the environment and the carrying capacity of the planet. Also, energy sustainability demands that enough energy should be available for the future generations. Equity in terms of energy is somewhat time-dependent, and it is expected that short-term differences will lessen over a period of time and energy opportunities in several countries will converge in the longer term.

- Lifestyles. The recent shift in lifestyles and tempering desires that are energy-driven can back the energy sustainability. As it is well known that the aspirations of people tend to increase continually, addressing this factor is often complicated. Transforming behavioral and decision-making patterns demands recognition that present development paths are not sustainable, but such recognition generally arises only when substantial short-term consequences are exhibited, e.g., disasters, oil-price shocks, and droughts.

To attain the investments required to lessen the risks related with energy initiatives, the public should take into consideration the probable long-term consequences associated with present behavior. Thus, the current and tough challenge for policy makers is turning future threats concerned with energy use into near-term priorities.

- Land use and aesthetics: There is need to balance the use of land for energy-related activities with other needs, such as recreation and agriculture. This tradeoff is a chiefly imperative challenge with technologies such as hydraulic energy that often embraces flooding large tracts of land, electricity transmission, which often traverses wind turbines, sensitive ecosystems. All these are highly transparent, and bioenergy should encourage the progress of energy plants on land that could be used for other purposes.

The governing principles of environment could affect the overall well-being of people, making it an important aspect of sustainability. It may be a challenging task to avert damage to environmental aesthetics e.g., renewable energy technologies such as wind turbines and solar collectors are considered by many to be detrimental to landscapes.

8.4. DRIVERS FOR RENEWABLE ENERGY POLICIES

Renewable energy technologies have the potential to provide energy services, including electricity and lighting, cooling, and heating, mobility, and mechanical energy. Moreover, comparative to other types of energy (from fossil fuels, traditional biomass, and nuclear power), modern renewables provide a range of further socio-economic benefits. In most jurisdictions, such socio-economic benefits lay critical role in driving policymakers to achieving renewable energy goal and support policies.

Renewable energy drivers (benefits) have been characterized into a number of ways. For instance, the United Nations Intergovernmental Panel on Climate Change (IPCC) Special Report on Renewable Energy Sources and Climate Change Mitigation (SRREN) classified vital drivers, prospects, and advantages of renewable energy into environmental (climate change alleviation and lessening of health and environmental impacts), access to energy, energy security (e.g., fuel imports; diversity of fuel supply; balance of trade), and economic and social development (e.g., rural development, job creation).

In this section, some of the key drivers in relation to the chief categories of economic, environmental, and political (security and social) criteria, recognizing that many other drivers can be categorized by more than one category.

8.4.1. Environmental Drivers

The extraction, conveyance, refining, and use of nuclear and fossil fuels result in a host of substantial environmental impacts, including harm to land from human deterioration activity such as mining; pollution of water and air; wastage of large amounts of freshwater, especially for cooling at power plants; danger of nuclear accidents; biodiversity loss; global climate change; and related impacts on the overall wellbeing of humans.

> For instance, the World Health Organization estimates that outdoor air pollution-due largely to the burning of coal and road transport-killed 3.7 million people worldwide in 2012. Another account estimates that 5.5 million people die prematurely each year due to household and outdoor air pollution; of that total, 1.6 million people die of air pollution in China and 1.4 million in India

The largest contributor to pollution in China is from coal-burning (that is the major cause for an estimated 366,000 deaths annually) while on the other hand, in India the primary contributor is the burning of solid biomass for heating and cooking. Biodiversity loss, health problems, and other environmental challenges will only worsen climate change conditions.

It was also observed that decisions related to the deployment of renewable energy have become an integral part of government strategies across the world to address many of these challenges.

8.4.2. Economic Drivers

Various technologies of renewable energy can provide a number of economic benefits, especially for energy importers. This is becoming progressively true as costs of renewable energy (particularly costs of wind power and solar PV) are consistently declining since the past few years. In addition, the wide use of renewable energy helps in averting a number of indirect economic costs concerned with the production of fossil energy and use, such as health care expenses.

It can also help in lessening the long term costs related with global climate change, such as the potential for sudden displacement and disruption of people and their economic activity (e.g., forced migration, spread of disease).

Thus, investing in renewable energy systems and related infrastructure facilities can help in attaining sustainable development goals in every sense of the term-sustained economic growth that is environmentally sustainable.

8.4.3. Political and Security Drivers

It is well known that energy is crucial for a healthy economy and energy services can sustain or enhance quality of life of people. Therefore, the continuous physical accessibility of reasonably economical and highly reliable energy products and services is increasingly important for political and to economic security. In a similar way, conditions of political security and economic stability are the essential foundation for the healthy growth of energy infrastructure and consistent supply of energy services.

Growing the proportion of energy share from indigenous renewables provide the opportunity to lessen reliance on imported fuels, improving security of supply. Renewables can help in enhancing the diversity of energy supply and contribute to resilience and flexibility of the energy system through local, distributed generation, both of which minimize the disruption impact of energy services.

The overall risk of prevalent technical failure of the energy system (for instance malfunction of system component) and risk of external physical disruption (terrorism, natural disasters, and piracy and sabotage) are potentially lessened by the use of disseminated renewables. It is because of the fact that each individual system component, (e.g., a wind turbine) becomes less critical to total system integrity.

8.5. MAIN POLICY STEPS FOR ENERGY SUSTAINABLE DEVELOPMENT

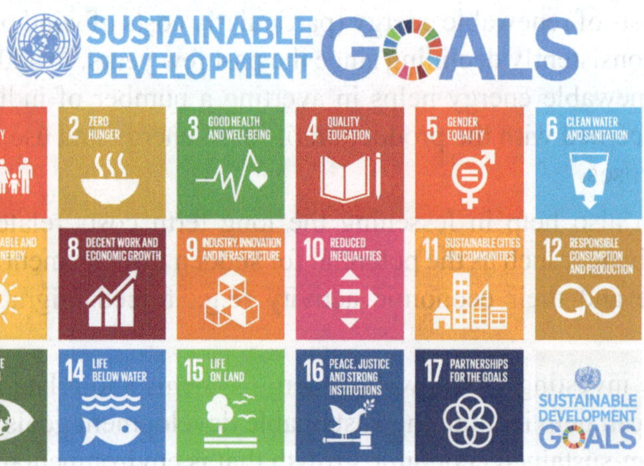

Figure 8.3. Sustainable development goals.

Sources: Image by Wikimedia Commons.

Reshaping policies to stimulates the transition to a sustainable energy system (Figure 8.3).

It is important to note that the objectives of energy sustainability are attainable, and need not contradict more short-term considerations, if the world embarks on a determined, collective effort. Reinventing the energy system to one in which

1. A systems perspective helps in designing and framing the overall policy; and
2. The transformation of energy from a variety of commodities to a series of services is not an easy task. At the outset, it is essential to start with the system that is in existence in the current time frame.

Nevertheless, several actions undertaken by the regulators and national governments, civil society and international organizations, and private sector investors will strengthen the necessary transformation. The details of the proposed actions can be found in the Yerevan Action Agenda and the Hammamet Declaration.

1. **Energy Markets Should Be Reformed:** in a way to make sure that the energy prices reflect full costs, including emissions, while eradicating subsidies of market-distorting throughout the system.

The use of energy subsidies could be lessened by exploring more effective and efficient ways to protect vulnerable groups. Policy-makers should work towards ensuring a changeover from an energy commodity industry to an energy services industry.

2. **Energy Market Reform Will Not Be Possible Unless Energy Security Strengthens:** A full range of normative instruments such as best practice guidance and standards is required throughout the energy system including development of international and regional norms covering interoperability, interconnections, and trading.

It is highly important to maintain an open dialogue among energy generating, conveying, and consuming countries on energy technology, security, and policy.

Ensuring mutually beneficial economic interdependence and greater interconnectivity demands high investment in energy infrastructure projects to augment energy efficiency, and optimize efficient utilization of energy resources and integration of renewable energy.

Encouraging interconnection infrastructure projects among countries with corresponding energy resources is a cost efficient method to boost mutual energy security.

3. **Energy Efficiency in Most Countries Needs to Improve More Quickly:** It is well known that enhancing energy efficiency is one of the most cost-effective options for fulfilling mounting energy demand in most countries, where there is high demand. It helps in providing better environment, contributing to energy security, enhanced life quality, and overall economic well-being.

Significant potential for enhancing efficiency of energy is in existence across the world, but efforts to enhance energy efficiency is always a challenging task because of faulty national policy frameworks:

- Policies that affectedly lower prices of energy and encourage wasteful consumption of critical resources; consumption and production subsidies distort markets;
- Housing stocks are regulated poorly;
- Inefficient land use management; entry barriers for new participants;
- There are inadequate standards and norms; and
- The information and statistics to manage energy use and track progress are incomplete.

Moreover, there is often a lack of education and public awareness about the long-term social and economic benefits of action to enhance industrial productivity and energy efficiency.

1. **Equitable Access to Modern Energy Services Requires Mobilizing Adequate Resources**: Ensuring economic and physical access to quality energy services necessitate the demand for the investment throughout the energy value chain, from chief energy development to end-use.

Enabling investment demands that the governments possess a long-term vision for delivering sustainable energy services, and that they circulate sustainable regulations and policies that allow consumers and producers to respond to a dynamically transitioning energy market.

Such a vision should have reliance on a total energy system perspective that involves provision of access to prevailing energy services for susceptible groups. It also demands fully integration of the full slate of development goals (e.g., gender, energy, and youth; the water-food-energy-ecosystems-health nexus).

2. **Renewable Energy Policies Need to be Redesigned:** It is generally seen that the renewable energy resources are consistently becoming cost-competitive in relation to conventional resources. They offer variety of option for lessening the overall net carbon intensity of the energy sector, enhance energy security, provide equitable energy access in remote areas, and facilitating economic development.

Integration of renewables into the global energy mix will play a very significant role as future energy systems are adjusted both on-and off-grid.

However, wider uptake of renewables requires finding solution to barriers in fair competition vis-à-vis conventional technology (without resorting to long-term subsidies), executing consistent long-term energy policy frameworks in a future energy system context, and arraying innovative and targeted financial mechanisms.

3. **Finance Will Be Critical:** It is usually seen that the transmission of the energy system requires mobilization of substantial financial resources. Therefore, it is of utmost important to align investment incentives with goal set by various governments for sustainable development.
4. **The Future Energy System will Need New Technology**: R&D as well as commercial launch of new technology, management skills and introduction of capital is indispensable to support the desirable transitions.

It plays a very important role as it helps in executing international collaboration on research and development of new technology as well as transfer of exchange lessons learned among member countries about large scale deployment of low carbon energy sources.

5. **Lies, Damned Lies, and Statistics:** It is very critical to develop appropriate indicators that illustrate evolution on energy for sustainable development goal of various countries. For instance, many countries have set deadlines to attain sustainable goals.

For this, there is a need on part of various countries to support energy statistics programs that govern and account for key variables of energy production and consumption, and that are fully integrated into other socio and economic national statistical efforts.

There is need to emphasize the need for collection of data on patterns of energy production and consumption and that should be in accordance with the future energy system needs and wants and to strengthen the analytical capacity of the distinctive interactions comprising energy policy in a way to provide innovative sustainable policy approaches to tackle multidisciplinary energy-related concerns.

8.6. POLICY OPTIONS FOR PROMOTING TECHNOLOGICAL INNOVATION

Figure 8.4. Conference for framing cleaner energy policies.

Source: Image by Wikimedia commons.

8.6.1. Research and Development

There are various options available to government to provide support to R&D. These include tax incentives to boost private R&D, grants, and contracts, and incentives, and schemes for collaborations between national and international players including combination of universities, private players and firms, and other research institutions (Figure 8.4).

8.6.2. Demonstration Projects

It is generally seen that putting demonstration efforts in order to prove technical viability at a near-commercial scale are costly, hard to finance as well as risky in nature. The public sector always been criticized for having poor track record in picking winners; its support for demonstration has barely resulted in successful commercial products.

The most efficacious demonstration efforts considered those in which the role of government role has been limited in setting cost goals and performance and providing some financial incentives and support. While on the other hand, private sector is fully responsible for their technological choices in achieving these goals and fiancé a major proportion of needed investment.

8.6.3. Buying Down Costs

For some radically latest energy technologies aimed at fulfilling sustainable development goals, there is need for some public subsidies and should be justifiable to facilitate private sector-led efforts aimed concerned with buying down technology costs.

There are many industrialized and some developing countries that offer financial packages to trigger new renewable energy development. The need for subsidies is of high importance especially where external costs are not fully internalized in energy prices. Public resources for technology cost buy-down should be apportioned with reference to criteria focused on augmenting societal returns on these scarce funds. Incentives should:

- Encourage lower technology costs through all the available means—including positive competition, economies in scale of production such as large scale or small scale, economies of learning through past experience, and gains from making gradual technological advancement as cumulative production grows.
- Have sunset provisions that limit their duration.
- Be limited to emerging technologies that provide major public assistances not fully valued in market prices, that have steep learning curves, and that have good future prospects for market penetration after the subsidies have been phased out.

To lessen and disseminate risks, buy-down programs can be proposed to support selections of new clean energy technologies. Initiatives coupled with positive competitive market transformation should be given full importance for two key reasons.

First, they have cost-savings potential. Second, the rate at which ongoing energy market is reforming provides an opportunity for introducing policy measures that would make it easier to acquire technology.

Some examples of competitive market transformation initiatives comprise Germany's former Feedstock Law, U.S. Renewable Portfolio Standard, and the recently discarded U.K. Non-Fossil Fuel Obligation.

8.6.4. Widespread Deployment

Even after successful acquire of technology new energy technologies aimed at fulfilling sustainable development goals often face imperative institutional obstacles to large scale deployment.

The removal of such hurdles—for instance, the fixing out of patent issues, or supports for domestic savings institutions—is an essential objective of public policy.

Many of the market reforms discussed in the section on making markets work better (internalizing externalities and removing subsidies) will accelerate widespread deployment of some new clean energy technologies.

These kinds of reforms will particularly be helpful in quickening the deployment of the new poly generation and cogeneration options. In addition to this, green energy marketing and energy labeling are made possible by market reforms that allow consumers to select their energy suppliers.

One critical element of policy issues relates to scale. Some bright new clean energy technologies are modular and small-scale, offering attractive economics through their ability for little cost through large scale production and their ability for deployment near users, where market values are high.

But employment of such technologies, that requires both small-scale production technologies and efficient end-use technologies, faces tremendous obstacles at both the producer and consumer levels.

8.7. CHALLENGES: ENERGY AND SUSTAINABLE DEVELOPMENT

Attaining sustainable development goals for energy is not an easy task. There are various challenges associated with the same. Some of the economic, political, and technical challenges are discussed as follows:

There are many countries that are consistently subsidizing fossil fuels as a way of minimizing the overall costs for consumers and encouraging economic growth. But subsidies that are not targeted directly are very costly and undermine efforts of energy conservation.

In the year 2013, it was noted that approximately $550 billion of public money was spent across the world on giving subsidies for these direct fossil fuel that are comparatively beneficial to wealthy people who use more than the poor.

In the month of January this year, the Indonesian Government took very determined decisions to implement a new fuel pricing system that has substantially reduced diesel and gasoline subsidy costs. This reduces fiscal risks and helps in allocating resources to more productive development projects.

Another significant economic challenge for majority of countries is that renewables are capital intensive and such hefty investment is very challenging to rise in risky environments. Many countries have executed policy incentives schemes to address these barriers. Countries such as India and Brazil-are having success with renewable energy auctions.

Some countries are also of the opinion that small-scale solar power can intensely accelerate energy access. Low-cost solar systems installed at homes helped countries such as Mongolia and Bangladesh to deliver energy services to low-income households who would otherwise be living in the dark.

According to a survey, it was noted that Bangladesh now has the largest national off-grid electrification program in the world. A project that was started in 2003, with connections for around 11,000 households, the program has now connected over 850,000 households to safe solar power every year.

And there are technical hurdles for the same: in case of large-scale, grid connected renewable energy; some of the major challenge is intermittency. Solar energy can only generate energy during the day and wind energy varies in intensity with respect to the current environmental conditions.

This makes it very difficult to integrate into the grid. But in countries such as Ireland and Denmark, these issues are being addressed too through micro-weather forecasting, complicated systems of transmission lines, and innovative grid management systems. They are showing the way forward for others.

Some of the major breakthroughs are now coming in the form of energy storage. New battery storage technology, especially those used lithium ion cells, could see the world transforming more rapidly to at least 50% renewable energy-or even greater.

8.8. THE WORLD BANK'S ROLE IN ENSURING SUSTAINABLE ENERGY DEVELOPMENT

Figure 8.5. Planning for sustainable development projects.

Source: Image by Pxfuel.

The role of the World Bank in ensuring sustainable development is important to take into consideration. It was noted that the World Bank Group take the issue of sustainable energy very seriously (Figure 8.5).

It is well known that getting there will entail tripling historic capital flows to access clean energy projects.

The World Bank plays a very important role in lending energy to low and middle-income countries. According to a survey, around 66% of energy lending goes to countries in Sub-Saharan Africa and South Asia where a large proportion of people live without any access to energy.

Almost 90% of power generation lending goes to cleaner energy forms: hydropower, natural gas, wind, solar, and geothermal. The World Bank does not make use of coal for power generation until there is any non-avoidable scenario.

World Bank plays a very crucial role in coordinating with countries to ensure a strong enabling environment that helps in encouraging greater investment-particularly in renewable energy sector.

The primary goal is to end energy poverty and ensure equitable distribution of energy to each and every citizen of nation.

It is widely believed that among the various potential sustainable development goals that receive overall global community acceptance, ending energy poverty will be topmost priority for this year.

Because of this, several experts of World Bank co-chair the Sustainable Energy for All initiative with the UN Secretary General. The key area of focus in on three goals: ensuring universal access to advance services of energy; quadrupling the production of renewable energy in the global energy mix; and doubling the rate of quality in energy efficiency.

In the present situation, all of us are experiencing negative impacts of climate change and pollution. But as students and energy experts of tomorrow, one can play a crucial role in delivering on the new energy sustainable development goal.

The world now need smart people who can come up and actively participate in achieving energy sustainable development goals. With contribution of each and every citizen of world, one can witness an end to energy poverty in our lifetimes.

8.9. THE CASE OF COMMUNITY ENERGY

Figure 8.6. How wind energy changed into electricity.

Source: Image by public domain pictures.

The appeal of local ownership through municipalization and energy cooperatives of local energy utilities has dramatically surge in the past few years. While the scenario of power cooperatives has long been in existence in Germany and elsewhere, going back over a century, there has been a noted comeback in the model over the past several years (Figure 8.6).

As of late 2013, it was noted that more than 800 cooperatives had been established in Germany alone in renewable energy sector, with the majority of them established between early 2009 and end-2012.

According to a survey, in 2012 citizen ownership stood at 47% of Germany's mounted renewable power capacity and 43% of its renewable power generation.

While, it is often seen that majority of the energy cooperatives emphasize on power generation, there are also latest cooperative district heating networks that provide connection of biogas plants to households, for instance, bringing together citizens, farmers, and the broader community to engage in decision making that directly impact all inhabitants as well as play crucial role in promoting local value.

Various researches has been done in Germany and other allied countries it was found that national interaction and participation plays a very crucial role in the public acceptance of renewable energy at all levels: this embrace the common viewpoint of renewable energy, as well as agreement of specific projects. Even so, the degree of agreement relied upon geographical location and technology.

There are several reasons for concerns to concrete local projects, varying from issues about aesthetics to possible impacts on native traffic patterns (for instance, delivery of biomass feedstock to new plants).

Concern about projects-as well as infrastructure, such as distribution lines and transmission-can be abridged by working to amplify individual assistances with context to the formation of public goods through procedural and distributive justice.

Distributive justice primarily emphasizes on who pays and who benefits-a sense of equitable distribution of benefits and costs positively affects acceptance of project.

Procedural justice is of the viewpoint that all affected have an equitable chance to contribute in the planning process and to be heard, even if their views do not rule the day. Research has basically categorized four key levels of participation: consultation, information, self-determination, and cooperation.Many policymakers and energy experts have long been considered involvement of general public to be significant for ensuring project acceptance. In Germany, the government and energy experts have recognized its significance for increasing acceptance.

For majority of the people, social acceptance of renewable energy projects is considered of high importance for a successful energy transition because of the certainly significant rise in renewable energy capacity, and because much of the energy production will occur nearby home.

The World Wind Energy Association acknowledges the importance of local community ownership which plays a very crucial role in higher social acceptance of wind power in Germany, Austria, the United States, and Scotland.

In a 2009 Scottish study, wind energy projects in two communities were examined, one where the project was executed by the national power company and the other (in a community with similar related conditions) where the wind farm privately owned by individuals.

Project acceptance (and decisions related to its expansion) was found to be higher in the latter because of higher profits generation for the municipality, the "green" image that developed from the project, and wider application of the project as a local success story.

In fact, the benefits generated from this will go far beyond the acceptance of renewable energy generally or of specific projects. Several studies concerning the significance of participation from local actors arrived at a conclusion that local ownership-as contrary to the usual land lease or other arrangements with traditional power producers-not only helps in strengthening the energy transformation but also benefits society as a whole.

Adopting a more individuals' centric strategy to energy production and use empowers farmers, citizens, and small businesses to invest in renewable energy projects, that help in creating profits in a community and is an imperative tool for local socio-economic development.

Citizen participation plays a very crucial role in building community relationship, while control and ownership over the means of energy production and consumption pattern can lead to a drop in total energy consumption. These studies have demonstrated several valuable effects beyond acceptance that are related to local participation and ownership in renewable energy projects, including:

- More-engaged citizenry-encouraged participation of community members to be engaged politically and socially, and development of superior competence in dealing with local government, thereby solidifying democracy;
- Increased awareness among general public and stakeholder of energy about its issues and importance of energy efficiency improvements;
- Increased probability of project triumph due to transparency and local participation in construction and planning activity,

particularly when project are designed by local people who are more aware of the ground level factors. High level of interaction is key for success of these projects;

- It is generally seen that stronger communities as projects bring people together, building a mutual identity, as well as pride in collective success and better feeling of self-worth among people involved;
- Reduced dependence on a restricted number of producers of energy and wider allocation of assets and encouragement within the energy system;
- Investing the money earned from energy in the community rather than its migration into global financial flows (through imports of electricity, fuels, etc.);
- Execution of projects-such as designing or framing networks of local heating or wind farms-that might not be developed by dominant players because of comparatively high transaction costs and project risk;

8.10. CONCLUSION

In the end, it is concluded that within the past few years, there has been increasing acceptance to achieve energy sustainable goals in various countries. The government in many counties has taken up initiatives and came up with several policies and strategies to find solution to energy problems.

The policies initiated by many countries should consider all the critical elements that are important for the success of the project as certain factors depend upon the local conditions which may vary from time to time.

There is need to significantly think about how to preserve the non-renewable energy as it is limited in nature and if it is not saved then it will be exhausted to use for the future generation.

REFERENCES

1. (2011). Energy for Sustainable Development in the UNECE Region. [e-book] Available at: https://www.unece.org/fileadmin/DAM/energy/se/pdfs/comm_gen/Publications/2017/UNECESustainableEnergyPub.pdf (accessed on 4 July 2020).
2. (2015). Energy Policies for Sustainable Development. [e-book] Available at: https://www.undp.org/content/dam/undp/library/Environment%20and%20Energy/Sustainable%20Energy/wea%202000/chapter12.pdf (accessed on 4 July 2020).
3. (2018). OECD Contribution to the United Nations Commission on Sustainable Development 15. [e-book] Organization for economic co-operation and development. Available at: https://www.oecd.org/greengrowth/38509686.pdf (accessed on 4 July 2020).
4. (2020). Renewable Energy and Sustainable Development. [e-book] World Future Council. Available at: https://www.worldfuturecouncil.org/wp-content/uploads/2016/08/WFC_2016_Renewable-Energy-and-Sustainable-Development.pdf (accessed on 4 July 2020).
5. Acciona.com, (2014). What is Sustainable Development and what are the Global Goals? | ACCIONA. [Online] Available at: https://www.acciona.com/sustainable-development/ (accessed on 4 July 2020).
6. Dincer, I., & Rosen, M., (1998). A worldwide perspective on energy, environment and sustainable development. International Journal of Energy Research, 22(15), 1305–1321.
7. Hendrickson, D., Corrigan, K., Keefe, A., Shaw, D., Jacob, S., Skelto, L., A. Schon, J., et al., (2020). Global Sustainability: An Authentic Context for Energy Education Journal of Sustainability Education. [online] Susted.com. Available at: http://www.susted.com/wordpress/content/global-sustainability-an-authentic-context-for-energy-education_2015_01/ (accessed on 4 July 2020).
8. Kienzl, (2020). Sustainability Trends: A Global Perspective-Design intelligence. [Online] DesignIntelligence. Available at: https://www.di.net/articles/sustainability-trends-global-perspective/ (accessed on 4 July 2020).
9. Nazam, A., & Cleveland, C. J., (2020). Energy and Sustainable Development at Global Environmental Summits. [Online] ResearchGate.

Available at: https://www.researchgate.net/publication/226292687_Energy_and_Sustainable_Development_at_Global_Environmental_Summits_An_Evolving_Agenda (accessed on 4 July 2020).

10. Rosen, M., (2009). Energy Sustainability: A Pragmatic Approach and Illustrations. Sustainability, 1(1), 55–80.
11. Sathaye, J., Lucon, O., & Rahman, A., (2012). Renewable Energy in the Context of Sustainable Development. [e-book] Available at: http://www.mcc-berlin.net/~creutzig/SRREN_Ch09.pdf (accessed on 4 July 2020).
12. Vezzoli, C., Ceschin, F., Osanjo, L., M'Rithaa, M. K., Moalosi, R., Nakazibwe, V., & Carel, D. J., (2020). Energy and Sustainable Development. [Online] SpringerLINK. Available at: https://link.springer.com/chapter/10.1007/978-3-319-70223-0_1 (accessed on 4 July 2020).
13. World Bank, (2020). Energy and Sustainable Development: What's Next? [Online] Available at: https://www.worldbank.org/en/news/speech/2015/06/10/energy-and-sustainable-development-whats-next (accessed on 4 July 2020).

INDEX

M

N

O

P

R

S

T

U

W